Lecture Notes in Mathematics

Edited by A. Dold, F. Takens and B. Teissier

Editorial Policy
for the publication of monographs

1. Lecture Notes aim to report new developments in all areas of mathematics – quickly, informally and at a high level. Monograph manuscripts should be reasonably self-contained and rounded off. Thus they may, and often will, present not only results of the author but also related work by other people. They may be based on specialized lecture courses. Furthermore, the manuscripts should provide sufficient motivation, examples and applications. This clearly distinguishes Lecture Notes from journal articles or technical reports which normally are very concise. Articles intended for a journal but too long to be accepted by most journals, usually do not have this "lecture notes" character. For similar reasons it is unusual for doctoral theses to be accepted for the Lecture Notes series.

2. Manuscripts should be submitted (preferably in duplicate) either to one of the series editors or to Springer-Verlag, Heidelberg. In general, manuscripts will be sent out to 2 external referees for evaluation. If a decision cannot yet be reached on the basis of the first 2 reports, further referees may be contacted: the author will be informed of this. A final decision to publish can be made only on the basis of the complete manuscript, however a refereeing process leading to a preliminary decision can be based on a pre-final or incomplete manuscript. The strict minimum amount of material that will be considered should include a detailed outline describing the planned contents of each chapter, a bibliography and several sample chapters.
Authors should be aware that incomplete or insufficiently close to final manuscripts almost always result in longer refereeing times and nevertheless unclear referees' recommendations, making further refereeing of a final draft necessary.
Authors should also be aware that parallel submission of their manuscript to another publisher while under consideration for LNM will in general lead to immediate rejection.

3. Manuscripts should in general be submitted in English.
Final manuscripts should contain at least 100 pages of mathematical text and should include
– a table of contents;
– an informative introduction, with adequate motivation and perhaps some
  historical remarks: it should be accessible to a reader not intimately familiar
  with the topic treated;
– a subject index: as a rule this is genuinely helpful for the reader.

Lecture Notes in Mathematics

1723

Editors:
A. Dold, Heidelberg
F. Takens, Groningen
B. Teissier, Paris

Springer
Berlin
Heidelberg
New York
Barcelona
Hong Kong
London
Milan
Paris
Singapore
Tokyo

Jean-Pierre Croisille   Gilles Lebeau

Diffraction by an Immersed Elastic Wedge

Springer

Author

Jean-Pierre Croisille
Laboratoire de Mathématiques
Université de Metz
57045 Metz, Cedex 01, France
E-mail: croisil@poncelet.univ-metz.fr

Gilles Lebeau
Centre de Mathématiques
École Polytechnique
91128 Palaiseau Cedex, France
E-mail: lebeau@math.polytechnique.fr

Cataloging-in-Publication Data applied for

Die Deutsche Bibliothek - CIP-Einheitsaufnahme

Croisille, Jean-Pierre:
Diffraction by an immersed elastic wedge / Jean-Pierre Croisille ;
Gilles Lebeau. - Berlin ; Heidelberg ; New York ; Barcelona ; Hong
Kong ; London ; Milan ; Paris ; Singapore ; Tokyo : Springer, 1999
   (Lecture notes in mathematics ; 1723)
   ISBN 3-540-66810-1

Mathematics Subject Classification (1991): Primary: 35G15, 35L20,
35L05,45F15, 76Q05, 78A45 Secondary: 30E20, 35J05, 78A40

ISSN 0075-8434
ISBN 3-540-66810-1 Springer-Verlag Berlin Heidelberg New York

© Springer-Verlag Berlin Heidelberg 1999
Printed in Germany

Typesetting: Camera-ready T_EX output by the authors
Printed on acid-free paper   SPIN: 10700343   41/3143-543210

Table of Contents

Acknowledgements. This work is greatly indebted to B. Poirée, physicist at the D.R.E.T. for his support, advice and his deep knowledge in acoustic. We gratefully acknowledge also F. Duclos (L.A.U.E., Université du Havre, URA 1373) and M. de Billy (G.P.S., Université Paris VI, UMR 7588) for many interesting discussions. We do not forget J. Laminie (Laboratoire d'Analyse Numérique, Université Paris XI, URA 760) for the support during the years of the numerical work.

1. Introduction

The study of wave diffraction by a wedge goes back to two articles by H. Poincaré published in 1892 and 1896 in Acta Mathematica [Po1, Po2]. His aim was to analyze in the harmonic regime the structure of the wave scattered by a perfectly conducting wedge in electromagnetism, and in particular the effect of the singularity on the polarization of this wave. The analytic solution of this problem was given by A. Sommerfeld in 1896, [Som1]. (See also [Som2])

In 1952, H.G. Garnir gave the Green function of the metaharmonic operator in a wedge, [Ga]. This work was generalized in 1958 by G.D. Maliuzhinets [Ma1,Ma2] to the more difficult case of impedance type boundary conditions on the faces of the wedge. These results were included by J.B. Keller in the sixties in his geometric theory of diffraction [K].

In 1982, J. Cheeger and M. Taylor [CT] proved the geometric localization of the singularities of the Green function of the wave equation in a manifold with conical singularities. In 1997, the second author [L3] proved the theorem of propagation of singularities for the wave equation with Dirichlet or Neumann conditions, in manifolds with curve- or wedge-type singular boundary. In the case of curved wedges, the asymptotics of the scattered wave was given in two dimensions by P. Gérard and G. Lebeau [GL], and by G. Lebeau [L1], in the case of wedges of codimension 2. Finally, J.M.L. Bernard [Be1,Be2] extended the method of Maliuzhinets in the case of an incident wave not perpendicular to the wedge. For an exhaustive study of wave diffraction by various shapes, we refer to [BSU]. See also [BM].

All these papers are devoted to the Maxwell or wave equations. One single propagation velocity is involved in the physics of the interaction with the wedge. The main topic of the present paper lies in the analysis of coupling by the wedge of different propagation velocities. In the special case which is studied, three types of waves with different velocities are present: the longitudinal and transversal waves in the elastic wedge and the acoustic sound wave in the fluid. In addition, two surface waves are present along the faces of the wedge. Firstly, the Scholte-Stoneley wave of the coupling fluid-solid by a nonsingular interface. Secondly, the Rayleigh wave of a free interface, which vanishes here, due to the presence of the fluid (i.e. becomes a resonance).

From a mathematical point of view, it is expected that the microlocalization of the coupling problem should allow to prove, as in [GL], that the asymptotics of the diffracted wave by a curved wedge is governed by the wave diffracted by the tangent wedge. Consequently, the main interest of the present work is to describe explicitly the *principal symbol* of the pseudo-differential operator connected with the coupling problem between the elastic wedge and the fluid. This symbol is called in the sequel the *spectral function* of the problem and is denoted by Σ. It is a vectorial ramified holomorphic function which is the unique solution in a convenient functional space of an integral equation with singular kernels. The structure of the singularities of this spectral function involves three noncommuting *translation operators* , disallowing the existence of an explicit representation formula like the one of Sommerfeld.

The other aspect of this work is the description of an efficient numerical algorithm for the approximation of the spectral function. This approximation is very accurate and allows, in particular, to compute diffraction diagrams in the high frequency limit. These diagrams have been compared successfully to experimental data provided by J. Duclos, A. Tinel, H. Duflo, [DTD, TD], in the case of an incident Scholte-Stoneley wave, [DTDL], and by M. de Billy, J.F. Piets [PB1,PB2] in the case of an incident wave in the fluid.

A detailed outline of the paper is given in Sect.2.2.4. Briefly, Sect.2 is devoted to the presentation of the problem and to the notation. The notion of outgoing solution is introduced and the definition of the spectral function is given. The main theoretical results (Theorem 1 and 2) about the existence, uniqueness and structure of the solution of the problem are given. The asymptotics of this solution in the far field is derived.

In Sect.3, the analytic structure of the spectral function is studied, especially its decomposition $\Sigma = y + X$, into a meromorphic part y and an holomorphic part X. Theorems 1 and 2 are proved in Sect.4. The Sect.5 and Sect.6 are devoted to the numerical study of the problem. The solution is approximated by a Galerkin-collocation method presented in Sect.5. Finally, we display in Sect.6 a broad series of numerical diffraction diagrams of an incident wave in the fluid or of Scholte-Stoneley type, for several angles of wedges, ranging from $150°$ to $25°$. Preliminary results of this work have been reported in [CL].

2. Notation and Results

2.1 Notation

We introduce in this section the notation used in the sequel of the paper. The angle of the wedge is φ, and we suppose $0 < \varphi < \pi$. We call Ω_s (resp. Ω_f) the angular section filled by the elastic medium (resp. the fluid)

$$
\begin{aligned}
\Omega_s &= \left\{ (x,y) = (r\cos\theta, r\sin\theta), \ 0 < \theta < \varphi, \ 0 < r \right\} \\
\Omega_f &= \left\{ (x,y) = (r\cos\theta, r\sin\theta), \ \varphi < \theta < 2\pi, \ 0 < r \right\}.
\end{aligned}
\tag{2.1}
$$

The common boundary is $\Gamma = \Gamma_1 \cup \Gamma_2$ where

$$
\Gamma_\alpha = \left\{ (x,y) = (r\cos\theta, r\sin\theta), \ \theta = 0 \ (\alpha = 1), \ \theta = \varphi \ (\alpha = 2) \right\}
\tag{2.2}
$$

are the two faces of the wedge (Fig. 2.1).

Moreover, we call

- $\mathbf{u}$: the vector displacement in the solid (the elastic medium).
- $\tilde{g} = g_{in} + g$: the potential of the velocity in the fluid, sum of the potential of the incident wave g_{in} and of the diffracted wave g.
- ρ_s : the density of the solid.
- ρ_f : the density of the fluid.
- λ, μ : the two Lamé parameters in the solid.
- $c_L = \sqrt{\frac{\lambda + 2\mu}{\rho_s}}$: the longitudinal velocity in the solid.
- $c_T = \sqrt{\frac{\mu}{\rho_s}}$: the transversal velocity in the solid.
- c_0 : the sound velocity in the fluid.
  We suppose that $c_0 < c_T < c_L$.
- c_s : the Scholte-Stoneley velocity at the interface fluid-solid (see Sect.3.2).

Under the hypothesis of small disturbances in the solid and the fluid (linearized equations, interface solid-fluid geometrically at rest), the constitutive equations are

$$\rho_s \partial_t^2 \mathbf{u} = (\lambda + \mu)\, \text{grad div}\, \mathbf{u} + \mu \Delta \mathbf{u} \qquad \text{in } \Omega_s \qquad (2.3)$$
$$\partial_t^2 \tilde{g} = c_0^2 \Delta \tilde{g} \qquad \text{in } \Omega_f \qquad (2.4)$$

with the two boundary conditions on Γ

$$\lambda (\text{div}\, \mathbf{u})\mathbf{n} + 2\mu \varepsilon(\mathbf{u}) \cdot \mathbf{n} = \rho_f \partial_t \tilde{g} \mathbf{n} \qquad (2.5)$$
$$\partial_t \mathbf{u} \cdot \mathbf{n} = \text{grad}\, \tilde{g} \cdot \mathbf{n} \qquad (2.6)$$

where $\mathbf{n}$ is the normal vector to the interface solid-fluid, and $\varepsilon(\mathbf{u})$ is the tensor of the deformations

$$\varepsilon_{ij}(\mathbf{u}) = \frac{1}{2}\left(\frac{\partial u_i}{\partial x_j} + \frac{\partial u_j}{\partial x_i}\right), \quad i,j = 1,2. \qquad (2.7)$$

The equation (2.5) expresses the continuity of the normal stress, ($\rho_f \partial_t \tilde{g}$ is the pressure in the fluid) and (2.6) expresses the continuity of the normal velocity. In this paper, we restrict ourself to time-harmonic solutions with frequency $f = \frac{\tau}{2\pi}$. The incident wave is given in the form

$$g_{in}(x,y,t) = k \; e^{i\tau t} e^{i\frac{\tau}{c_0}(x \cos\theta_{in} - y \sin\theta_{in})} \qquad (2.8)$$

where θ_{in} is the angle between the face 1 of the wedge and the direction of the incident wave, and k is a constant.

The dimensionless form of the problem is obtained by defining the functions v and h by

$$\mathbf{u}(x,y,t) = e^{i\tau t} v\left(\frac{\tau}{c_L}x, \frac{\tau}{c_L}y\right) \qquad (2.9)$$
$$g(x,y,t) = e^{i\tau t} c_L h\left(\frac{\tau}{c_L}x, \frac{\tau}{c_L}y\right) \qquad (2.10)$$

The corresponding dimensionless constants are

$$\begin{cases} \mu = \dfrac{\mu}{\rho_s c_L^2}; \; \lambda = \dfrac{\lambda}{\rho_s c_L^2} \\[2mm] \rho = \dfrac{\rho_f}{\rho_s} \\[2mm] \nu_0 = \dfrac{c_L}{c_0}; \;\; \nu_T = \dfrac{c_L}{c_T}; \;\; \nu_L = 1; \;\; \nu_S = \dfrac{c_L}{c_S}. \end{cases} \qquad (2.11)$$

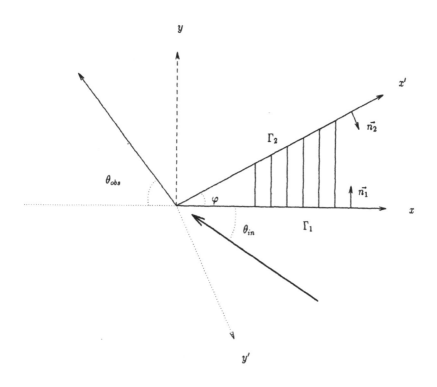

Fig. 2.1. The wedge of angle φ, illuminated by a plane wave in the fluid

Since $c_L = \sqrt{\frac{\lambda+2\mu}{\rho_s}}$, $c_T = \sqrt{\frac{\mu}{\rho_s}}$, we have $\lambda+2\mu = 1$ and $\mu = \frac{1}{\nu_T^2}$. Moreover, because we have supposed $c_T > c_0$, we have $\nu_L = 1 < \nu_T < \nu_0$. Denoting the dimensionless elasticity operator by $E = (\lambda + \mu)\operatorname{grad}\operatorname{div} + \mu\Delta$, the system (2.3-2.6) is equivalent to the equations (2.12-2.15)

$$
\begin{cases}
(2.12) & (E+1)v & = 0 & \text{in } \Omega_s \\
(2.13) & (\Delta + \nu_0^2)h & = 0 & \text{in } \Omega_f \\
(2.14) & (\lambda\operatorname{div} v + 2\mu\varepsilon(v))\mathbf{n} - i\rho h\mathbf{n} & = i\rho h_{in}\mathbf{n} & \text{on } \Gamma \\
(2.15) & iv\cdot\mathbf{n} - \operatorname{grad} h\cdot\mathbf{n} & = \operatorname{grad} h_{in}\cdot\mathbf{n} & \text{on } \Gamma
\end{cases}
\quad (S)
$$

The incident dimensionless plane wave is taken as

$$h_{in}(x,y) = \frac{1}{2} \, e^{i\nu_0(x\cos\theta_{in} - y\sin\theta_{in})}. \tag{2.16}$$

In Sect.2.3 we shall explain the meaning of an outgoing solution for the equations (2.12)-(2.15). The existence and uniqueness of an outgoing solution to our problem is stated in Theorem 1, Sect.2.5. We see by the formulas (2.9), (2.10), that the high frequency asymptotics of the displacement $\mathbf{u}(x,y)$ and of the potential $g(x,y)$ is given by the behaviour of $v(x,y)$ and $h(x,y)$ when (x,y) runs to infinity along a ray. In order to keep the symmetry between the two faces of the wedge, we introduce two orthogonal frames $(0,x,y)$ and $(0,x',y')$ respectively related to the faces of the wedge (cf Fig. 2.1)

$$\begin{cases} \overrightarrow{OM} = x\mathbf{t} + y\mathbf{n}, & \mathbf{t} = (1,0), \ \mathbf{n} = (0,1) \\ \overrightarrow{OM} = x'\mathbf{t}' + y'\mathbf{n}', & \mathbf{t}' = (\cos\varphi, \sin\varphi), \ \mathbf{n}' = (\sin\varphi, -\cos\varphi). \end{cases} \tag{2.17}$$

2.2 Strategy of the Study

We describe hereafter the main steps of the study, and we introduce the key notions used in the sequel.

2.2.1 Decomposition of the Solution and Spectral Function

The basic idea of the study is to seek a solution of the problem as the sum of two contributions, one for each face of the wedge. Let us consider for example the wave equation $(\Delta + \nu_0^2)h = 0$ in the plane $\mathbb{R}^2$ cut by the half axis $D = \{(x,y)/(x \geq 0, y = 0)\}$. If $\gamma(x)$ is a distribution supported on D, then the distribution

$$h(x,y) = (\Delta + \nu_0^2)_+^{-1}[\gamma(x)\delta(x \geq 0, y = 0)] \tag{2.18}$$

is a solution of this equation. Defining $h(x,y)$ as the superposition $h(x,y) = h_1(x,y) + h_2(x',y')$, where h_1 corresponds to $D = \Gamma_1$ and h_2 to $D = \Gamma_2$, allows to define a solution of the wave equation in $\mathbb{R}^2 \setminus (\Gamma_1 \cup \Gamma_2)$. Similarly, we look for $v(x,y)$, solution of the elasticity equation $(E+1)v = 0$ in the form $v(x,y) = v_1(x,y) + v_2(x',y')$, where the term $v_j(x',y')$ is solution of the elasticity operator in $\mathbb{R}^2 \setminus \Gamma_j$.

In (2.18), the subscript "+" indicates that the correct *outgoing* solution is selected. This is done by defining (2.18) as the limit when $\varepsilon \to 0^+$ of the distribution

$$h^{\varepsilon}(x, y) = (\Delta + \nu_0^2 e^{-2i\varepsilon})^{-1}[\gamma(x)\delta(x \geq 0, y = 0)]$$

Note that the outgoing solution cannot be selected by a radiation condition, like the one of Sommerfeld, because the wedge Ω_s is an unbounded domain. In fact, for an incidence angle $\theta_{in} \in]0, \frac{\pi}{2}[$, the outgoing solution contains a reflected part against the face 1, at every time in the past, having an $O(1)$ amplitude with respect to the distance to the origin.

After some manipulations, it can be proved that the couple (v_1, h_1) has the form of a Fourier integral like

$$\begin{bmatrix} v_1^x(x, y) \\ v_1^y(x, y) \\ h_1(x, y) \end{bmatrix} = \int_{\mathbb{R}} S(\xi, y) \cdot \begin{bmatrix} \hat{\alpha}_1(\xi) \\ \hat{\beta}_1(\xi) \\ \hat{\gamma}_1(\xi) \end{bmatrix} e^{ix\xi} d\xi$$

where $S(\xi, y)$ is a 3x3 complex matrix and $\Sigma_1(\xi) = \left[\hat{\alpha}_1(\xi), \hat{\beta}_1(\xi), \hat{\gamma}_1(\xi)\right]^T$ is the Fourier transform of layer potentials supported by Γ_1. The parameter $\xi \in \mathbb{C}$ is the Fourier parameter of the face 1. By replacing in this formula the variables (x', y') by the variables (x, y), we obtain $[v_2^{x'}, v_2^{y'}, h_2]^T$ in function of $\Sigma_2(\xi) = [\hat{\alpha}_2(\xi), \hat{\beta}_2(\xi), \hat{\gamma}_2(\xi)]^T$. The couple of functions $\Sigma(\xi) = (\Sigma_1(\xi), \Sigma_2(\xi))$ is called in the sequel, *the spectral function* of the problem and is the main tool of this study. In fact, the high frequency asymptotics of the diffracted wave in the fluid is directly related by a stationnary phase theorem to the value of the spectral function on the segment $[-\nu_0, \nu_0]$.

2.2.2 Structure of the Spectral Function and Physical Interpretation

In order to solve the coupled problem (S), we have now to translate in terms of $\Sigma_1(\xi)$ and $\Sigma_2(\xi)$ the boundary conditions (2.14)-(2.15). These conditions are proved to be equivalent to an integral system of the form

$$\begin{cases} DM \cdot \Sigma_1 + TM \cdot \Sigma_2 = S_1(\xi) \\ TM \cdot \Sigma_1 + DM \cdot \Sigma_2 = S_2(\xi) \end{cases} \tag{2.19}$$

In this system, $DM(\xi, \zeta)$ and $TM(\xi, \zeta)$ are singular integral kernels, each one being a 3×3 matrix.

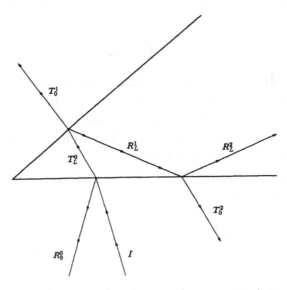

Fig. 2.2 Incident (I), reflected ($R_{L,0}^i$) and transmitted ($T_{L,0}^i$) rays

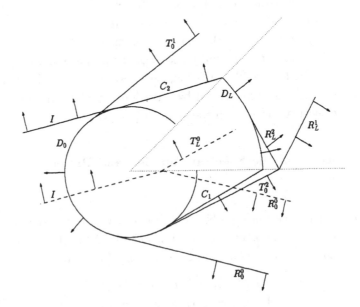

Fig. 2.3 The wave fronts at two times $-t_0 < 0$ (dashed), and $4t_0 > 0$ (continuous)

The kernel $DM(\xi, \zeta)$ corresponds to the coupling solid/fluid by each face of the wedge. The kernel $TM(\xi, \zeta)$ corresponds to the interaction of the waves between the two faces of the wedge. These operators are described in details in Sect.3.1-3.3. The first (respectiveley second) equation in (2.19) corresponds to the boundary conditions on the first (respectiveley second) face. The right hand side describes the incident wave, seen by the first (repectiveley second) face. Because it is basically Fourier transform of functions with support on $\mathbb{R}^+$, the spectral function is holomorphic in the lower half plane $\mathrm{Im}\,\xi < 0$. Moreover, the numerical computation as well as the existence and uniqueness of (Σ_1, Σ_2) rely upon the explicit description of its poles in the upper half-plane.

Roughly speaking, this description is possible, due to the fact that the integral system (2.19) can be written formally

$$\begin{cases} \Sigma_1 - T \cdot \Sigma_2 = S_1'(\xi) \\ -T \cdot \Sigma_1 + \Sigma_2 = S_2'(\xi) \end{cases}$$

where the operator T is

$$T = -DM^{-1}.TM$$

and the r.h.s. is $(S_1', S_2') = (DM^{-1}S_1, DM^{-1}S_2)$. Therefore, we have a representation of Σ_1, Σ_2 in the form of Neumann series

$$\Sigma_1 = (\sum_{p \text{ even}} T^p).S_1' + (\sum_{q \text{ odd}} T^q).S_2' + R_1' \qquad \Sigma_2 = (\sum_{p \text{ even}} T^p).S_2' + (\sum_{q \text{ odd}} T^q).S_1' + R_2'$$

where $R_{1,2}'$ is a remainder. Each term $T^n.S'$, $(S' = S_{1,2}')$ has the form

$$T^n.S'(\xi) = P_n(\xi) + R_n(\xi)$$

where $P_n(\xi) = \sum_j \frac{w_j}{\xi - Z_{n,j}}$, $\mathrm{Im}\,Z_{n,j} > 0$, $w_j \in \mathbb{C}^3$, and $R_n(\xi)$ is a regular function. This leads to a decomposition of (Σ_1, Σ_2) in the form

$$\Sigma_j(\xi) = y_j(\xi) + X_j(\xi)$$

where y_1, y_2 are two meromorphic functions corresponding to the P_i, having their poles in the upper half-plane, and X_1, X_2 are two regular functions corresponding to the R_i.

The physical meaning of the functions y_1, y_2 is as follows. Each function $\xi \to \frac{w}{\xi - Z}, w \in \mathbb{C}^3, \mathrm{Im}\,Z \geq 0$ occuring in $y_j(\xi)$ corresponds in fact to an *incoming* plane wave on the face j. Suppose that the incident wave in the fluid is seen only by the face 1, $(0 < \theta_{in} < \pi - \varphi/2)$. This wave is the function

$h_{in}(x,y) = \frac{1}{2} e^{i\nu_0(x\cos\theta_{in} - y\sin\theta_{in})}$. Its counterpart in $y_1(\xi)$ is a function proportional to $\xi \to \frac{1}{\xi - \nu_0\cos\theta_{in}}$. By refraction at the interface 1, it generates inside the wedge two plane waves, of type L and T. They are real waves if $|\nu_0\cos\theta_{in}| \leq \nu_{L,T}$, and vanishing waves if $|\nu_0\cos\theta_{in}| \geq \nu_{L,T}$. These two waves are incident on the face 2 and correpond in $y_2(\xi)$ to two functions proportional to $\xi \to \frac{1}{\xi - \nu_*\cos(\theta_* + \varphi)}$, $* = L$ or T, where the complex angle θ_* is defined by $\nu_0\cos\theta_{in} = \nu_*\cos\theta_*$. Each one generates in turn by reflection on the face 2, two additional waves in the wedge (L and T) incident on the face 1, giving a new contribution in y_1. This mechanism defines a recurrence process permitting to build two sequences of poles ordered in generations. The smallest is the angle φ of the wedge, the largest is the number of poles occuring in $y_1(\xi)$, $y_2(\xi)$. Subtracting $y_j(\xi)$ from $\Sigma_j(\xi)$ gives a remainder function called the X−part of the spectral function. This X−part is an holomorphic function in the domain $\mathbb{C}\backslash] - \infty, -1]$. No explicit representation formula is available for this part. However, it can be computed very accurately by a Galerkin-collocation method. This method has been carefully worked out (§5) and numerically tested (§6) on a broad series of cases of physical interest. It can be seen as the own contribution of the vertex of the wedge to the solution.

We display on Fig 2.2 the geometric construction of the rays occuring in the problem. For the sake of simplicity, only one type of rays is pictured inside the wedge, (type L). We note on Fig. 2.2, 2.3 by R the reflected waves, and by T the transmitted waves. The subscript indicates the type of wave (L in the solid, 0 in the fluid). The superscript indicates the generation number of the pole.

On Fig 2.3 is pictured the wave fronts configuration at two times. The dashed lines correspond to the incoming wave at time $-t_0 < 0$ before it reaches the edge. The continuous lines correspond to the fronts at time $4t_0$. The velocity is two times faster in the wedge than in the fluid. The diffracted fronts are D_0 in the fluid, D_L in the solid. The fronts corresponding to the so-called critical angles of the two faces of the wedge are C_1, C_2.

In Sect.2.7, we give the asymptotics of the diffracted field D_0 off the transmitted and reflected fronts. The analysis of the asymptotics in the neighborhood of these two fronts amounts to study the stationnary phase theorem at the points where the integrand does possess a pole. This analysis is not carried out in the sequel. In fact, in the physical experiments, the base time of the oscilloscopes is accurate enough to distinguish the arrival times of the different waves, so that the diffraction diagrams computed in Sect.6 are directly comparable to experimental diagrams.

2.2.3 The Recursive Equation

The structure of the singular integral system (2.19) of which $\Sigma(\xi)$ is solution allows us to deduce an important equation satisfied by the spectral function, called the *recursive equation* (Sect.3.5). It looks like

$$\Sigma(\xi) = g(\xi) + M_L(\xi)\Sigma(\xi_L) + M_T(\xi)\Sigma(\xi_T) + M_0(\xi)\Sigma(\xi_0). \qquad (2.20)$$

In this equation, $g(\xi)$ is a well behaved function and the matrices $M_*(\xi)$, $* \in \{L, T, 0\}$, are known explicitly. The point ξ is in the lower half plane and ξ_*, $* \in \{L, T, 0\}$, is given by the inverse of the translation T_*, (Sect.2.4). If $\xi = \nu_* \cos\theta$, then $\xi_* = \nu_* \cos(\theta - \varphi)$. Thus, ξ_* is obtained by a "rotation" along an ellipse of ξ from the left to the right in the lower half-plane. The equation (2.20) plays an important role in the sequel. One of its interests is the following. Because of the numerical method used for the computation of $X(\xi)$, the spectral function is more accurately computed in the right lower quadrant than in the left one. Thus, using (2.20) recursively permits to propagate into the left quadrant the accuracy of the computation of the right one. This method has been proved to be very efficient, especially for small values of the angle φ of the wedge (§5, §6).

2.2.4 Outline

The rest of Section 2 is devoted to the presentation of the two main theoretical results. After introducing the notion of outgoing solutions for the system (S) in Sect.2.3; we define in Sect.2.4 the translation operators connected to each type of waves (L−wave and T−wave in the solid, sound wave indexed by (0) in the fluid). In Sect.2.5, we give the existence and uniqueness theorem for the outgoing waves (Theorem 1), and the structure theorem for the spectral function (Theorem 2). The computation of the asymptotics of the diffracted wave in the fluid is derived in Sect.2.7.

Section 3 is entirely devoted to the study of the spectral function. The singular kernel system to solve is derived in Sect.3.1. In Sect.3.2, we study the interface wave solution of the coupled problem elastic medium-fluid by a plane interface. We recover at this stage the existence of the well-known *Scholte-Stoneley* wave. In Sect.3.3, we prove various analytical and algebraic properties of the integral system, allowing to exhibit in the Sect.3.4 the decomposition of the spectral function into two parts, the one meromorphic, the other holomorphic. Finally, we introduce in Sect.3.5 the recursive equation satisfied by the spectral function, which plays an important role in the sequel, both from analytical and numerical points of view.

In section 4, we prove the theoretical results of the paper. The first one (Sect.4.1) is an isomorphism theorem, used as an auxiliary result for Theorems 1 and 2. In Sect.4.2, Sect.4.3 these two theorems are proved.

The two last sections are devoted to the numerical aspects of the study. Firstly, we describe in section 5 the different tools used for the numerical computation of the spectral function. Two cases of physical interest have been studied. Firstly, the one of an incident plane wave in the fluid. Secondly, the one of an incident Scholte-Stoneley wave, along the first face of the wedge. In each case, we describe carefully the numerical algorithm for the computation of the meromorphic part (y−part) of the spectral function (§5.2, §5.3). Then, we describe in Sect.5.4, Sect.5.5 the method of Galerkin-collocation used for the numerical approximation of the holomorphic part (X−part). Finally, we show in the Sect.5.6 how the recursive equation (Sect.2.2.3, Sect.3.5), can be used as a useful tool for the computation of the spectral function in regions of the complex plane where the Galerkin-collocation method is useless.

In the last section (Sect.6), we present a broad series of numerical results for the diffraction by a dural wedge immersed in water. Diffraction diagrams are displayed and commented, for angles of the wedge ranging from 150° to 25°. Even at the latter angle, the method is stable and gives very accurate results.

2.3 Outgoing Solutions

In this paragraph, we define the notion of an *outgoing solution* for the system (S). For this purpose, we shall consider these equations as the special case when $\varepsilon = 0$ of the following system where $\varepsilon \in [0, \pi[$ of the system (S_ε)

$$
\begin{cases}
(E + e^{-2i\varepsilon})v^\varepsilon = 0 \text{ in } \Omega_s \\
(\Delta + \nu_0^2 e^{-2i\varepsilon})h^\varepsilon = 0 \text{ in } \Omega_f \\
(\lambda \operatorname{div} v^\varepsilon + 2\mu\varepsilon(v^\varepsilon))\mathbf{n} - ie^{-i\varepsilon}\rho h^\varepsilon \mathbf{n} = ie^{-i\varepsilon}\rho h_{in}\mathbf{n} \text{ on } \Gamma \\
(ie^{-i\varepsilon}v^\varepsilon \cdot \mathbf{n} - \operatorname{grad} h^\varepsilon \cdot \mathbf{n}) = \operatorname{grad} h_{in} \cdot \mathbf{n} \text{ on } \Gamma.
\end{cases}
\quad (2.21) \ (S_\varepsilon)
$$

Let $f \in S'(\mathbb{R})$ be a tempered distribution with support in $\mathbb{R}_+$. Its Fourier transform

$$\hat{f}(\xi) = \int_{\mathbb{R}} e^{-ix\xi} f(x) dx \qquad (2.22)$$

is holomorphic in the lower half-plane $\operatorname{Im} \xi < 0$.

Definition 2.1 (Definition of the class $\mathcal{A}$). *We say that $f \in \mathcal{A}$ if $f \in \mathcal{S}'(\mathbb{R})$, $supp(f) \subset \mathbb{R}^+$ and if*

$$\exists C_0 > 0 \ \text{such that} \quad \sup_{-\pi < \theta < 0} \int_{\rho > C_0} |\hat{f}(\rho e^{i\theta})|^2 d\rho < +\infty \qquad (\mathcal{A}.1)$$

$\hat{f}(\xi)$ is holomorphic in the neighborhood of the points

$$\xi = \nu_L, \ \xi = \nu_T, \ \xi = \nu_0, \ \xi = \nu_S \qquad (\mathcal{A}.2)$$

We call $\hat{\mathcal{A}}$, the set of the Fourier transforms of functions belonging to $\mathcal{A}$. Denoting by $\mathcal{F}$ the Fourier transform on $\mathbb{R}^2$, we have for any $v \in \mathcal{S}'(\mathbb{R}^2)^2$, $h \in \mathcal{S}'(\mathbb{R}^2)$

$$\mathcal{F}(Ev) = -M\mathcal{F}(v); \quad \mathcal{F}(\Delta h) = -(\xi^2 + \eta^2)\mathcal{F}(h) \qquad (2.23)$$

where M is the elasticity symbol

$$M(\xi, \eta) = \begin{bmatrix} (\lambda + \mu)\xi^2 + \mu(\xi^2 + \eta^2) & (\lambda + \mu)\xi\eta \\ (\lambda + \mu)\xi\eta & (\lambda + \mu)\eta^2 + \mu(\xi^2 + \eta^2) \end{bmatrix}. \qquad (2.24)$$

The eigenvectors and eigenvalues of $M(\xi, \eta)$ are, for $(\xi, \eta) \neq (0, 0)$.

$$\begin{cases} M \begin{bmatrix} \xi \\ \eta \end{bmatrix} = (\xi^2 + \eta^2) \begin{bmatrix} \xi \\ \eta \end{bmatrix} & L - \text{mode (longitudinal)} \\ M \begin{bmatrix} -\eta \\ \xi \end{bmatrix} = \dfrac{\xi^2 + \eta^2}{\nu_T^2} \begin{bmatrix} -\eta \\ \xi \end{bmatrix} & T - \text{mode (transversal).} \end{cases} \qquad (2.25)$$

For $\varepsilon \in]0, \pi[$, the operators $E + e^{-2i\varepsilon}$ and $\Delta + e^{-2i\varepsilon}\nu_0^2$ are uniformly elliptic and are bijective onto $\mathcal{S}'(\mathbb{R}^2)$. If δ_1 and δ_2 are the integration measures on the faces F_1 and F_2 of the wedge

$$\langle \delta_1, g \rangle = \int_0^{+\infty} g(r, 0) dr; \quad \langle \delta_2, g \rangle = \int_0^{+\infty} g(r \cos \varphi, r \sin \varphi) dr \qquad (2.26)$$

we have the following lemma (the proof is postponed to the end of §2.6).

Lemma 2.2. *Let $\alpha_j, \beta_j, \gamma_j \in \mathcal{A}$, $j = 1, 2$. Then the two tempered distributions v_j^ε, h_j^ε defined for $\varepsilon \in]0, \pi[$ by*

$$\begin{cases} v_j^\varepsilon = -(E + e^{-2i\varepsilon})^{-1} \left[\begin{pmatrix} \alpha_j \\ \beta_j \end{pmatrix} \otimes \delta_j \right] \\ h_j^\varepsilon = -(\Delta + \nu_0^2 e^{-2i\varepsilon})^{-1} [\gamma_j \otimes \delta_j] \end{cases} \tag{2.27}$$

converge when $\varepsilon \to 0$ to two distributions of $\mathcal{S}'(\mathbb{R}^2)$ v_j^0, h_j^0 solutions of

$$\begin{cases} (E + 1)v_j^0 = - \begin{pmatrix} \alpha_j \\ \beta_j \end{pmatrix} \otimes \delta_j \\ (\Delta + \nu_0^2)h_j^0 = - \gamma_j \otimes \delta_j. \end{cases} \tag{2.28}$$

Moreover the following regularity results hold for any $\varepsilon \in [0, \pi[$:

(i) v_j^ε, h_j^ε are continuous on $\mathbb{R}^2$ and the traces v_j^ε / Γ, h_j^ε / Γ are in the space $H^1_{\text{loc}}(\Gamma)$,

(ii) the traces $\partial_n^+ v_j^\varepsilon / \Gamma$, $\partial_n^- h_j^\varepsilon / \Gamma$ do exist in the space $L^2_{\text{loc}}(\Gamma)$, and are tempered distributions in a neighborhood of $r = +\infty$.

Definition 2.3. *(Outgoing solutions). We call outgoing solution (v, h) of the equations (2.12)-(2.15) a solution (v, h) of the form*

$$v = v_1{}_{|\Omega_s} + v_2{}_{|\Omega_s}, \quad h = h_1{}_{|\Omega_f} + h_2{}_{|\Omega_f}$$

with

$$v_j = - \lim_{\varepsilon \to 0}(E + e^{-2i\varepsilon})^{-1} \left[\begin{pmatrix} \alpha_j \\ \beta_j \end{pmatrix} \otimes \delta_j \right]$$

$$h_j = - \lim_{\varepsilon \to 0}(\Delta + \nu_0^2 e^{-2i\varepsilon})^{-1} [\gamma_j \otimes \delta_j]$$

where $\alpha_j, \beta_j, \gamma_j \in \mathcal{A}$, $j = 1, 2$.

2.4 Translation Operators

An important tool in the sequel is the complex cosine transformation defined by

$$\cos(\theta_1 + i\theta_2) = \cos\theta_1 \cosh\theta_2 - i\sin\theta_1 \sinh\theta_2. \tag{2.29}$$

on the domain $\overset{\circ}{\mathcal{D}} = \{\theta \in \mathbb{C}, 0 < \operatorname{Re}\theta < \pi\}$. This function is a biholomorphic map between $\overset{\circ}{\mathcal{D}}$ and the domain $\Omega = \{z \in \mathbb{C} \backslash \{z \in \mathbb{R}, |z| \geq 1\}\}$. For θ_2 fixed, the open segment $]0, \pi[+i\theta_2$ oriented from left to right is mapped onto the half ellipse with focus ± 1 and semi-axis $\cosh\theta_2$, $|\sinh\theta_2|$ oriented from right to left. (The segment $]0, \pi[$ is mapped onto $]-1,1[$). Similarly, the vertical axis $\mathrm{Re}\,\theta = \theta_1 \in]0; \pi[$ with orientation from top to bottom is mapped onto the hyperbolic branch with focus ± 1 and orientation from bottom to top.

We extend the cosine function to a bijection from $\mathcal{D} = \overset{\circ}{\mathcal{D}} \cup \{\theta = -it, t \geq 0\} \cup \{\theta = \pi + it, t \geq 0\}$ onto the whole complex plane $\mathbb{C} = \Omega \cup] - \infty, -1] \cup [1, +\infty[$. For $* \in \{L, T, 0\}$ we introduce now the function defined on $\mathcal{D}$ by $\theta \mapsto \nu_* \cos\theta$. This function has the previous properties of the cosine function, except that the points ± 1 are replaced by the points $\pm\nu_*$. We introduce also the function $\zeta_*(\xi)$ defined for $\xi \in \mathbb{C}$ by the relation

$$\zeta_*(\xi) = -\nu_* \sin\theta, \quad \xi = \nu_* \cos\theta, \quad \theta \in \mathcal{D} \tag{2.30}$$

and which is such that $\zeta_*(\xi) = -\nu_*(\sin\theta_1 \cosh\theta_2 + i\cos\theta_1 \sinh\theta_2)$. For $\xi \in \mathbb{R}$, $\zeta_*(\xi)$ is related to the standard square root by the relations

$$\begin{cases} \zeta_*(\xi) = -\sqrt{\nu_*^2 - \xi^2} & \text{if } |\xi| \leq \nu_* \\ \zeta_*(\xi) = i\sqrt{\xi^2 - \nu_*^2-} & \text{if } |\xi| > \nu_*. \end{cases} \tag{2.31}$$

We consider now the domain Ω_* defined by (cf. Fig. 2.5)

$$\Omega_* = \left\{ \xi = \nu_* \cos\theta, \ \theta \in \mathcal{D}, \ \mathrm{Re}\,\theta < \pi - \varphi \right\} \tag{2.32}$$

Ω_* is the region of the complex plane on the right of the hyperbola $\xi = \nu_* \cos\theta, \mathrm{Re}\,\theta = \pi - \varphi$. For $\xi \in \Omega_*$, we define the *translation operator* T_*

$$T_*(\xi) = \nu_* \cos(\theta + \varphi) = \cos\varphi \ \xi + \sin\varphi \ \zeta_*(\xi). \tag{2.33}$$

More explanations on the map $T_*(\xi)$ will be given in §3. If $\xi_1, \xi_2 \in \mathbb{C}$ are given, we introduce now two sets of complex values $Z^1(\xi_1, \xi_2)$, $Z^2(\xi_1, \xi_2)$ defined by

$$Z^1(\xi_1, \xi_2) = \bigcup_{\ell \geq 0} Z^1_\ell, \quad Z^2(\xi_1, \xi_2) = \bigcup_{\ell \geq 0} Z^2_\ell. \tag{2.34}$$

Z^1_ℓ, Z^2_ℓ are called *generation number* ℓ, and are constructed by the following recurrence relation

$$\begin{cases} Z^1_0 = \xi_1, \ Z^2_0 = \xi_2 \\ Z^1_{\ell+1} = \tilde{Z}^1_{\ell+1} \cup T_0(\tilde{Z}^2_\ell \cap \Omega_0), \ \ell \geq 0 \\ Z^2_{\ell+1} = \tilde{Z}^2_{\ell+1} \cup T_0(\tilde{Z}^1_\ell \cap \Omega_0), \ \ell \geq 0 \end{cases} \tag{2.35}$$

where the auxiliary sets $\tilde{Z}^1_\ell$, $\tilde{Z}^2_\ell$, are defined by

$$\begin{cases} \tilde{Z}_0^1 = \xi_1, \ \tilde{Z}_0^2 = \xi_2 \\ \tilde{Z}_{\ell+1}^1 = T_L(\tilde{Z}_\ell^2 \cap \Omega_L) \cup T_T(\tilde{Z}_\ell^2 \cap \Omega_T), \ \ell \geq 0 \\ \tilde{Z}_{\ell+1}^2 = T_L(\tilde{Z}_\ell^2 \cap \Omega_L) \cup T_T(\tilde{Z}_\ell^1 \cap \Omega_T), \ \ell \geq 0. \end{cases} \quad (2.36)$$

Lemma 2.4. *The sets* Z^1, Z^2 *are finite.*

Proof. It is sufficient to prove that an infinite sequence z_ℓ of the form $z_0 \in \mathbb{C}$, $z_{\ell+1} = T_{\nu_\ell}(z_\ell)$, $z_\ell \in \Omega_{\nu_\ell}$, $\nu_\ell \in \{\nu_L, \nu_T, \nu_0\}$ does not exist. If $z = \nu \cos\theta = \nu(\cos\theta_1 \cosh\theta_2 - i \sin\theta_1 \sinh\theta_2)$, $\theta \in \mathcal{D}$, $0 \leq \text{Re}\,\theta < \pi - \varphi$ then $\text{Re}(T_\nu(z)) = \nu \cos(\theta_1 + \varphi) \cosh\theta_2 \leq \nu(\cos\theta_1 - \varepsilon_0) \cosh\theta_2 \leq \text{Re}\,z - \varepsilon_0\nu$, where $\varepsilon_0 > 0$ depends only of φ. Thus, if the number of points z_ℓ is infinite, then $\text{Re}\,z_\ell \to -\infty$ and $|z_\ell| \to +\infty$. This insures that $|\theta_2^\ell| \to +\infty$ and $\cosh\theta_2^\ell \sim |\sinh\theta_2^\ell| \sim \frac{1}{2}e^{\theta_2^\ell}$. Consequently, $|\text{Arg}\,z_\ell| \sim \theta_\ell^1 \in]0, \pi - \varphi[$ and $|\text{Arg}\,z_{\ell+1}| \sim \theta_\ell^1 + \varphi \sim \theta_{\ell+1}^1$. This gives a contradiction since it implies $\theta_\ell^1 \to +\infty$. $\blacksquare$

We refer to Sect.2.2.2. for the physical interpretation of the sets Z^1, Z^2.

2.5 Main Theorems

The first result concerns the existence and unicity of outgoing solutions of the problem, as settled in Definition 2.3. We need to suppose that the incidence angle θ_{in} is such that

$$\{\nu_L, \nu_T, \nu_0\} \cap Z^j(\nu_0 \cos\theta_{in}, \nu_0 \cos(\theta_{in} + \varphi)) = \emptyset \qquad j = 1, 2 \qquad \text{(H)}$$

The hypothesis (H) means that the incident wave

• is not an incomming grazing wave along one of the two faces.

• does not generate through the recurrence formula (2.35) an incomming grazing wave of the fluid-solid coupling by a plane interface. Because of Lemma 2.4, the hypothesis (H) is satisfied except for a finite set of incidence angles θ_{in}.

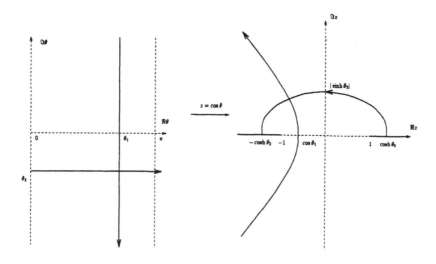

Fig. 2.4 The complex cosine transformation between the domainsc $\overset{\circ}{\mathcal{D}}$ and Ω.

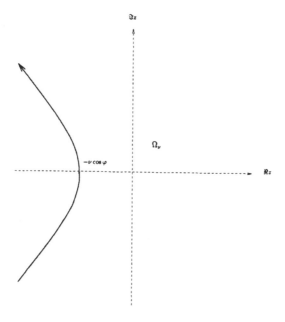

Fig. 2.5. The domain Ω_ν

Theorem 1. *(Existence and uniqueness of an outgoing solution)*

Under the hypothesis (H), the system (2.12)-(2.15) does have an unique outgoing solution (v, h).

If $\alpha_j, \beta_j, \gamma_j \in \mathcal{A}$ are the single layer potentials linked to the outgoing solution (v, h) by (2.28), we define the *spectral function* of the system (5) by

$$\Sigma_j(\xi) = \begin{bmatrix} \hat{\alpha}_j(\xi) \\ \hat{\beta}_j(\xi) \\ \hat{\gamma}_j(\xi) \end{bmatrix}, \quad j = 1, 2. \tag{2.37}$$

The proof of Theorem 1 is postponed to Sect.4.2.

We denote by $\mathbb{U}$ the domain

$$\mathbb{U} = \mathbb{C}\backslash]-\infty, -\nu_L] \tag{2.38}$$

and by $\mathcal{H}$ the space of the functions f, holomorphic on $\mathbb{U}$ and such that $f(xe^{i\alpha}) \in L^2(\mathbb{R}, dx)$ for each $\alpha \in]0, \pi[$. Moreover, for $j = 1, 2$ we note

$$\mathcal{P}^j = Z^j\left(\nu_0 \cos\theta_{in}, \nu_0 \cos(\theta_{in} + \varphi)\right) \tag{2.39}$$

$$\mathcal{C}^j = Z^j(-\nu_L, -\nu_L) \cup Z^j(-\nu_T, -\nu_T) \cup \{-\nu_0\}. \tag{2.40}$$

The set $\mathcal{C}^j$ is such that $\mathcal{C}^j \subset [-\nu_0, -\nu_L]$. Recall also that c_S is the Scholte-Stoneley velocity and $\nu_S = \frac{c_L}{c_S}$ (cf. §3.2). The following theorem describes the structure of the spectral function.

Theorem 2. *(Structure of the spectral function)*

• *For $j = 1, 2$, the function Σ^j is meromorphic on $\mathbb{U}$. More precisely, we have the decomposition*

$$\Sigma_j(\xi) = y_j(\xi) + X_j(\xi), \quad j = 1, 2 \tag{2.41}$$

where $X_j \in \mathcal{H}$ and $y_j(\xi) = \sum\limits_m \frac{w_j^m}{\xi - z_j^m}$, $z_j^m \in \mathcal{P}^j \cap \mathbb{U}$.

• *The boundary values $\Sigma_j(\xi - i0)$, $\xi \in \mathbb{R}$ are analytical for $\xi \notin \mathcal{P}^j \cup \mathcal{C}^j \cup \{-\nu_S\}$. Moreover, under the generic hypothesis $\mathcal{P}^j \cap \mathcal{C}^j = \emptyset$, the function $\Sigma_j(\xi - i0)$ has only simple poles in the set $(\mathcal{P}^j \cap \mathbb{R}) \cup \{-\nu_S\}$. Finally, the following equality holds in a neighborhood of $\xi_0 \in \mathcal{C}^j$*

$$\Sigma_j(\xi - i0^+) = a(\xi) + (\xi - \xi_0)^{1/2} b(\xi) \tag{2.42}$$

where $a(\xi), b(\xi)$ are holomorphic in a neighborhood ξ_0.

The proof of Theorem 2 is postponed to the Sect. 4.3.

2.6 Integral Representation of the Solution

The integral form of an outgoing solution of the system (2.12-2.15) will directly result from the proof of Lemma 2.2. Let us detail now this proof. We limit ourself to prove the result for $v_1 = v$. The proofs for the functions h_1, v_2, h_2 are similar. Let $\zeta_*^\varepsilon(\xi)$ be the root with positive imaginary part of the equation

$$\xi^2 + \zeta_*^\varepsilon(\xi)^2 = \nu_*^2 e^{-2i\varepsilon}. \tag{2.43}$$

The function $\zeta_*^\varepsilon(\xi)$ can be rewritten as $\zeta_*^\varepsilon(\xi) = e^{-i\varepsilon}\zeta_*(e^{i\varepsilon}\xi)$ where ζ_* is given by (2.30). The matrix $[M(\xi,\eta) - e^{-2i\varepsilon}]^{-1}$ is meromorphic in $\eta \in \mathbb{C}$ and its poles are located in the sets $K_\pm^\varepsilon(\xi)$ defined by $K_+^\varepsilon(\xi) = \{\zeta_L^\varepsilon(\xi), \zeta_T^\varepsilon(\xi)\}$, $K_-^\varepsilon(\xi) = \{-\zeta_L^\varepsilon(\xi), -\zeta_T^\varepsilon(\xi)\}$. By definition of v^ε (cf. 2.27) we have

$$\widehat{v^\varepsilon}(\xi,\eta) = [M(\xi,\eta) - e^{-2i\varepsilon}]^{-1} \begin{bmatrix} \hat\alpha(\xi) \\ \hat\beta(\xi) \end{bmatrix} \tag{2.44}$$

with $\alpha, \beta \in \mathcal{A}$. We introduce, for $y \in \mathbb{R}$, $\xi \in \mathbb{R}$ the matrix $L^\varepsilon(y,\xi)$

$$L^\varepsilon(y,\xi) = \int_\mathbb{R} e^{iy\eta}[M(\xi,\eta) - e^{-2i\varepsilon}]^{-1} d\eta \tag{2.45}$$

$L^\varepsilon(y,\xi)$ is continuous in y, analytical in ξ and can be rewritten for $y > 0$ and $y < 0$ in the form

$$L_{\pm y>0}^\varepsilon = \int_{\gamma_\pm(\xi)} e^{iy\eta}[M(\xi,\eta) - e^{-2i\varepsilon}]^{-1} d\eta \tag{2.46}$$

where $\gamma_+(\xi)$ (resp. $\gamma_-(\xi)$) is a loop enclosing $K_+^\varepsilon(\xi)$ in the direct sense (resp. $K_-^\varepsilon(\xi)$ in the inverse sense). Since the matrix M is homogeneous of order 2 in (ξ,η) and the functions $\zeta_*^\varepsilon(\xi)$ converge uniformly with respect to $\xi \in \mathbb{R}$ to $\zeta_*(\xi)$, we deduce from the inequality

$$\exists c_0 > 0/|\xi| \geq \nu_T + 1 \Rightarrow \text{Im}\,\zeta_*(\xi) \geq 2c_0|\xi| \tag{2.47}$$

that the matrices $L_{\pm y>0}^\varepsilon$ converge to the matrix $L_\pm^0$, defined by (2.45) with $\varepsilon = 0$. The convergence holds in the space $C^\infty(\pm y > 0)$, (depending on ξ), $|\xi| \geq \nu_T + 1$, equipped with the semi-norms

$$\sup_{\pm y>0,\xi} |\partial_y^k L(y,\xi)| e^{c_0|y||\xi|}|\xi|^{1-k} < +\infty. \tag{2.48}$$

Moreover, we have for $|\xi| \geq \nu_T + 1$, $L_+^0(y,\xi) = L_-^0(y,\xi)$. We choose now $M \geq \nu_T + 1$ such that $1_{|\xi|\geq M-1}(\hat\alpha(\xi), \hat\beta(\xi)) \in L^2$, and we decompose v^ε into two parts $v^\varepsilon = v_1^\varepsilon + v_2^\varepsilon$, studied separately

$$v_1^\varepsilon(x,y) = \frac{1}{4\pi^2} \int_{|\xi|\geq M} e^{ix\xi} L^\varepsilon(y,\xi) \begin{bmatrix} \hat{\alpha}(\xi) \\ \hat{\beta}(\xi) \end{bmatrix} d\xi \tag{2.49}$$

$$v_2^\varepsilon(x,y) = \frac{1}{4\pi^2} \int_{\gamma_\varepsilon} e^{ix\xi} L^\varepsilon(y,\xi) \begin{bmatrix} \hat{\alpha}(\xi) \\ \hat{\beta}(\xi) \end{bmatrix} d\xi \tag{2.50}$$

where γ_ε is pictured on Fig. 2.6.

It follows from (2.48) that $v_1^\varepsilon \in C^0(y, H_x^1)$ and converges in this space to v_1^0. Therefore $v_1^0(x,y)$ is continuous. The affirmations (i), (ii) of Lemma 2.2 for the part v_1^0 result easily from (2.48). We have for example

$$(\partial_y v_1^0)(r\cos\varphi, r\sin\varphi)_{r>0} = \frac{1}{4\pi^2} \int_{|\xi|\geq M} e^{ir\cos\varphi\xi} [\partial_y L_+^0(r\sin\varphi, \xi)] \begin{bmatrix} \hat{\alpha}(\xi) \\ \hat{\beta}(\xi) \end{bmatrix} d\xi. \tag{2.51}$$

This function is in the space $L^2(r > 0)$, because of $(2.48)_{k=1}$ and the fact that $f \mapsto \int_0^{+\infty} e^{-uv} f(v) dv$ is bounded on $L^2(\mathbb{R}_+)$.

For the study of the part $v_2^\varepsilon(x,y)$, we need to compute the matrix $L^\varepsilon(y,\xi)$. For any $(\xi, \eta) \neq (0,0)$, the decomposition of the vector $\begin{bmatrix} \hat{\alpha} \\ \hat{\beta} \end{bmatrix}$ onto the eigenvectors of $M(\xi, \eta)$ is

$$\begin{bmatrix} \hat{\alpha} \\ \hat{\beta} \end{bmatrix} = \psi_L \begin{bmatrix} \xi \\ \eta \end{bmatrix} + \psi_T \begin{bmatrix} -\eta \\ \xi \end{bmatrix} \Leftrightarrow \psi_L = \frac{\xi\hat{\alpha} + \eta\hat{\beta}}{\xi^2 + \eta^2}; \quad \psi_T = \frac{-\eta\hat{\alpha} + \xi\hat{\beta}}{\xi^2 + \eta^2} \tag{2.52}$$

We deduce from (2.52) that

$$[M - e^{-2i\varepsilon}]^{-1} \begin{bmatrix} \hat{\alpha} \\ \hat{\beta} \end{bmatrix} = \frac{\nu_L^2 \psi_L}{\xi^2 + \eta^2 - \nu_L^2 e^{-2i\varepsilon}} \begin{bmatrix} \xi \\ \eta \end{bmatrix} + \frac{\nu_T^2 \psi_T}{\xi^2 + \eta^2 - \nu_T^2 e^{-2i\varepsilon}} \begin{bmatrix} -\eta \\ \xi \end{bmatrix} \tag{2.53}$$

Applying the Cauchy formula to (2.46) we get

$$L_{\pm y>0}^\varepsilon \begin{bmatrix} \hat{\alpha} \\ \hat{\beta} \end{bmatrix} = i\pi e^{2i\varepsilon} \left[e^{i|y|\zeta_L^\varepsilon} \left(\pm\hat{\beta} + \frac{\xi}{\zeta_L^\varepsilon}\hat{\alpha} \right) \begin{bmatrix} \xi \\ \pm\zeta_L^\varepsilon \end{bmatrix} \right.$$
$$\left. + e^{i|y|\zeta_T^\varepsilon} \left(\mp\hat{\alpha} + \frac{\xi}{\zeta_T^\varepsilon}\hat{\beta} \right) \begin{bmatrix} \mp\zeta_T^\varepsilon \\ \xi \end{bmatrix} \right] \tag{2.54}$$

Because of $\operatorname{Im}\zeta_*^\varepsilon(\xi) > 0$ (cf. (2.43)), the branch points of the function $\zeta_*^\varepsilon(\xi)$ are $\pm e^{-i\varepsilon}\nu_*$. Moreover, since $\hat{\alpha}, \hat{\beta}$ are holomorphic in a neighborhood of ν_L, ν_T, we can replace in (2.50) the path γ_ε by a path $\tilde{\gamma}$, independent of ε. The functions

$$v_{2,\pm}^\varepsilon(x,y) = v_{2|\pm y>0}^\varepsilon = \frac{1}{4\pi^2} \int_{\tilde{\gamma}} e^{ix\xi} L_{|\pm y>0}^\varepsilon \begin{bmatrix} \hat{\alpha} \\ \hat{\beta} \end{bmatrix} d\xi \tag{2.55}$$

are (see (2.54)) analytical in $x \in \mathbb{R}$, $\pm y > 0$, coincide on $y = 0$ and converge uniformly towards

$$v_{2,\pm}^0 = \frac{1}{4\pi^2} \int_{\tilde{\gamma}} e^{ix\xi} L_\pm^0 \begin{bmatrix} \hat{\alpha} \\ \hat{\beta} \end{bmatrix} d\xi. \tag{2.56}$$

The properties (i), (ii) of Lemma 2.2 for v_2^0 result from the analyticity of $v_{2,\pm}^0$ with respect to $x \in \mathbb{R}$, $\pm y > 0$ and their coincidence on $y = 0$. We deduce from the preceeding results that the displacement $v_1^\varepsilon(x,y)$ is given, for $0 \le \varepsilon < \pi$, by the integral formula

$$\begin{aligned}
v_1^\varepsilon(x,y)_{|\pm y>0} &= \frac{i}{4\pi} e^{2i\varepsilon} \int_{\Gamma_0} e^{ix\xi} \left(e^{i|y|\zeta_L^\varepsilon} \left(\pm \hat{\beta}_1 + \frac{\xi}{\zeta_L^\varepsilon} \hat{\alpha}_1 \right) \begin{bmatrix} \xi \\ \pm \zeta_L^\varepsilon \end{bmatrix} \right. \\
&\quad \left. + e^{i|y|\zeta_T^\varepsilon} \left(\mp \hat{\alpha}_1 + \frac{\xi}{\zeta_T^\varepsilon} \hat{\beta}_1 \right) \begin{bmatrix} \mp \zeta_T^\varepsilon \\ \xi \end{bmatrix} \right) d\xi
\end{aligned} \tag{2.57}$$

where Γ_0 is the path pictured on Fig. 2.7. The same computation gives the integral formula for h_1^ε, $0 \le \varepsilon < \pi$

$$h_1^\varepsilon(x,y) = \frac{i}{4\pi} \int_{\Gamma_0} e^{i(x\xi+|y|\zeta_0^\varepsilon)} \frac{\hat{\gamma}_1(\xi)}{\zeta_0^\varepsilon} d\xi. \tag{2.58}$$

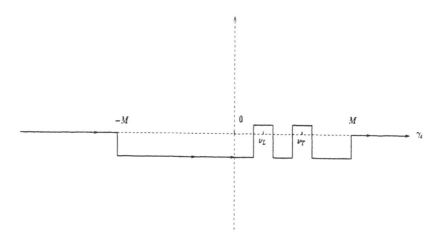

Fig. 2.6 The contour γ_ε

The formulas for $(v_2^\varepsilon, h_2^\varepsilon)$ (functions of the face 2) are similar, with the substitution of (x,y) by (x',y')

$$\begin{aligned}
v_2^\varepsilon(x',y')_{|\pm y'>0} &= \frac{i}{4\pi} e^{2i\varepsilon} \int_{\Gamma_0} e^{ix'\xi} \left(e^{i|y'|\zeta_L^\varepsilon} \left(\pm \hat{\beta}_2 + \frac{\xi}{\zeta_L^\varepsilon} \hat{\alpha}_2 \right) \begin{bmatrix} \xi \\ \pm \zeta_L^\varepsilon \end{bmatrix} \right. \\
&\quad \left. + e^{i|y'|\zeta_T^\varepsilon} \left(\mp \hat{\alpha}_2 + \frac{\xi}{\zeta_T^\varepsilon} \hat{\beta}_2 \right) \begin{bmatrix} \mp \zeta_T^\varepsilon \\ \xi \end{bmatrix} \right) d\xi
\end{aligned} \tag{2.59}$$

$$h_2^\varepsilon(x',y') = \frac{i}{4\pi} \int_{\Gamma_0} e^{i(x'\xi+|y'|\zeta_0^\varepsilon)} \frac{\hat{\gamma}_2(\xi)}{\zeta_0^\varepsilon} d\xi. \qquad (2.60)$$

In the preceeding formulas, the choice of the root $\zeta_*^0(\xi)$ is obtained by analytical continuation along Γ_0, with $\zeta_*^0(z) = \zeta_*(z)$ for z near 0. We have still for $z \in \Gamma_0 \cap \mathbb{R}$,

$$\zeta_*^0(z) = \zeta_*(z) = \begin{cases} i\sqrt{z^2 - \nu_*^2} & , \ |z| \geq \nu_* \\ -\sqrt{\nu_*^2 - z^2} & , \ |z| < \nu_* \end{cases} \qquad (2.61)$$

This allows to replace in the formulas (2.57)-(2.60) the path Γ_0 by the real axis and the function $\zeta_*^0(z)$ by $\zeta_*(z)$. The products $\zeta_*^{-1/2}(z)\hat{f}(z)$ are well defined for $f \in \mathcal{A}$; the fact that the traces are tempered distributions follows.

For the simplicity of the notation, we shall still call Γ_0 in the sequel, the real axis with contour of $-\nu_*$ from below and of ν_* from above, for each value of the symbol $*$ in the set $\{L, T, 0\}$.

2.7 Asymptotics of the Diffracted Wave in the Fluid

Let $M = (x,y) = (-R\cos\theta, -R\sin\theta)$ be an observation point located in the fluid. The number N of wavelengths between M and 0 is

$$N = \frac{\tau R}{2\pi c_0} = f\frac{R}{c_0}. \qquad (2.62)$$

We are interested in the behavior of the diffracted field g in the fluid for large values of N (far field regime), and we shall take $\rho = 2\pi N$ as a large parameter. Thanks to the scaling formula (2.10), we have to evaluate the asymptotics

$$h\left(-\frac{\rho\cos\theta}{\nu_0}, -\frac{\rho\sin\theta}{\nu_0}\right) \qquad \rho \to +\infty. \qquad (2.63)$$

To do so, we use the decomposition $h = h_1 + h_2$ where h_1 and h_2 are defined by formulas (2.58), (2.60) with $\varepsilon = 0$. The change of variables $\xi = \nu_0\cos\beta$, $d\xi = -\nu_0\sin\beta \ d\beta = \zeta_0(\nu_0\cos\beta)d\beta$ leads to , by denoting $h_1 = h_1\left(-\frac{\rho\cos\theta}{\nu_0}, -\frac{\rho\sin\theta}{\nu_0}\right)$, and $h_2 = h_2\left(-\frac{\rho\cos\theta}{\nu_0}, -\frac{\rho\sin\theta}{\nu_0}\right)$

$$\begin{cases} h_1 = \frac{i}{4\pi} \int_{C_0} e^{-i\rho[\cos\theta\,\cos\beta+|\sin\theta|\sin\beta]}\hat{\gamma}_1(\nu_0\cos\beta)d\beta \\ \\ h_2 = \frac{i}{4\pi} \int_{C_0} e^{-i\rho[\cos(\theta-\varphi)\cos\beta+|\sin(\theta-\varphi)|\sin\beta]}\hat{\gamma}_2(\nu_0\cos\beta)d\beta \end{cases} \qquad (2.64)$$

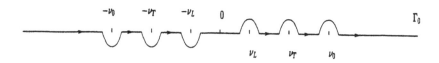

Fig. 2.7. The contour Γ_0

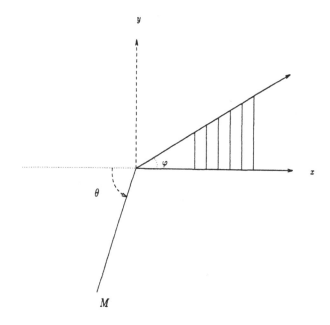

Fig. 2.8. An observation point M under angle θ

where C_0 is the complex path in the β-plane, pictured on Fig 2.9. The fact that $\hat{\gamma}_1(\nu_0\cos \beta)$ is an even function of β allows to eliminate the absolute value $|\sin \theta|$ in formula (2.64) by writing

$$\begin{cases} h_1 = \dfrac{i}{4\pi} \displaystyle\int_{C_0} e^{-i\rho \cos (\theta-\beta)} \hat{\gamma}_1(\nu_0\cos \beta) d\beta & \text{if } \theta \geq 0 \\[2ex] h_1 = \dfrac{i}{4\pi} \displaystyle\int_{C_1} e^{-i\rho \cos (\theta-\beta)} \hat{\gamma}_1(\nu_0\cos \beta) d\beta & \text{if } \theta \leq 0 \end{cases} \qquad (2.65)$$

where C_1 is the complex path pictured on Fig. 2.10. We can rewrite (2.65) in the form

$$h_1 = \frac{i}{4\pi} \int_{\gamma_\theta} e^{-i\rho \cos (\theta-\beta)} \hat{\gamma}_1(\nu_0\cos \beta) d\beta \qquad (2.66)$$

where γ_θ is the complex path pictured on Fig. 2.11. We have $\mathrm{Im}[\cos(\theta - \beta)] < 0$ in the shadowed regions. By Theorem 2, we know that the function $\hat{\gamma}_1(\nu_0\cos \beta)$ is holomorphic near $\beta = 0$, and under the generic hypothesis $\mathcal{P}^j \cap \mathcal{C}^j = \emptyset$, has only a finite number of simple poles and ramifications on the segment $[0,\theta]$. We can therefore under this hypothesis separate in (2.66) the contribution of the critical point $\beta = \theta$ and of the singularities of $\hat{\gamma}_1(\nu_0\cos \beta)$ in $[0,\theta]$, for regular values of θ, by writing

$$h_1 = h_1^D + h_1^S \qquad (2.67)$$

where h_1^D is the contribution of the stationary point

$$h_1^D \simeq -\frac{i}{4\pi} e^{-i\rho} e^{i\pi/4} \sqrt{\frac{2\pi}{\rho}} \hat{\gamma}_1(\nu_0\cos \theta) \qquad (2.68)$$

(we have used the classical stationary phase expansion at first order), and where h_1^S denotes the contribution to the expansion of h_1 of the singularities of $\hat{\gamma}_1(\nu_0\cos \beta)$ in $]0,\theta[$. Notice that the h_1^S part contains only a finite number of plane waves (due to reemission in the fluid of multiple refracted waves in the solid, and to reemission at the critical angles of the coupling fluid-solid). This corresponds on Fig. 2.3 to the contribution of the fronts R, T (poles) and to the critical fronts C_j (ramification points).

In particular, if one rotates a transducer in the fluid, at a fixed distance from 0, and if this transducer is sufficiently directional, the measured signal will be approximately proportional to

$$|\hat{\gamma}_1(\nu_0\cos \theta) + \hat{\gamma}_2(\nu_0\cos(\theta - \varphi))| \qquad (2.69)$$

for θ not too close to the finite set of singularities of the function $\hat{\gamma}_1, \hat{\gamma}_2$.

Our numerical algorithm (Sect.5) allows us to evaluate the boundary values of the functions $\hat{\gamma}_j(\xi - i0^+)$ on the real axis, and we have compared with good agreements the diffraction diagram given by (2.69) to experimental results, (Sect.6).

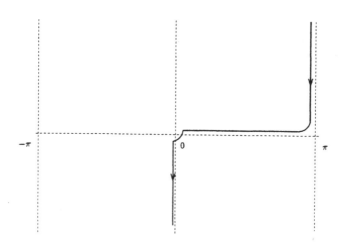

Fig. 2.9 The path C_0

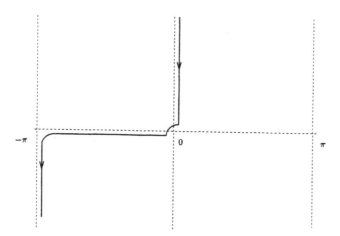

Fig. 2.10 The path C_1

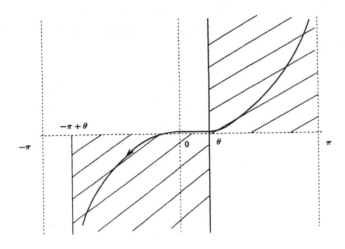

Fig. 2.11 The path γ_θ, $\theta \geq 0$

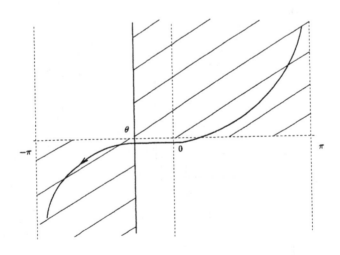

Fig. 2.12 The path γ_θ, $\theta \leq 0$

3. The Spectral Function

3.1 The Integral System for the Spectral Function

In the preceding section, we have described the class of the so-called outgoing solutions (v, h) for the system (2.12-2.15). They are of the form

$$v = v_1 + v_2; \quad h = h_1 + h_2 \tag{3.1}$$

each pair of functions (v_j, h_j), $j = 1, 2$ being given by the values at $\theta = 0$ of the distributions v_j^θ, h_j^θ, (see Definition 2.3).

$$v_j^\theta = -(E + e^{-2i\theta})^{-1} \left[\begin{pmatrix} \alpha_j \\ \beta_j \end{pmatrix} \otimes \delta_j \right]; \quad h_j^\theta = -[\Delta + v_0^2 e^{-2i\theta}]^{-1} [\gamma_j \otimes \delta_j] \tag{3.2}$$

where $\alpha_j, \beta_j, \gamma_j \in \mathcal{A}$. The distributions v_j^θ, h_j^θ are known explicitly by the integral representation formulas (2.57-2.60). In addition, these formulas are still valid for $\theta = 0$ (cf Sect.2.6). We call $\Sigma = (\Sigma_1, \Sigma_2)$, where $\Sigma_j = [\hat{\alpha}_j, \hat{\beta}_j, \hat{\gamma}_j]^T$, the spectral function occuring in (3.2). The aim of this section is to translate on Σ the boundary conditions (2.14-2.15). Because of the proof given in Sect.2.6, the boundary estimates (i), (ii) of Lemma 2.2 are still valid for v_j^θ, h_j^θ.

Let us introduce the boundary operator A^θ, $\theta \in [0, \pi[$ for the face 1 (left hand side of (2.21)(iii)-(iv)).

$$A^\theta(\Sigma_1, \Sigma_2) = \begin{cases} (\lambda \operatorname{div} v^\theta + 2\mu\varepsilon(v^\theta)) \cdot \mathbf{n} - ie^{-i\theta} \rho h^\theta \mathbf{n}|_{\Gamma_1} \\ ie^{-i\theta} v^\theta \cdot \mathbf{n} - \operatorname{grad} h^\theta \cdot \mathbf{n}|_{\Gamma_1} \end{cases} \tag{3.3}$$

$A^\theta(\Sigma_1, \Sigma_2)$ is a function of $x \geq 0$ with values in $\mathbb{C}^3$, which lies in the space $L^2_{\text{loc}}(x \geq 0)$ and is a tempered distribution in the neighborhood of $+\infty$.

Lemma 3.1. *For $\xi \in \mathbb{C}$, $\mathrm{Im}\,\xi < 0$, the Fourier transform of the boundary operator $A^\theta(\Sigma_1, \Sigma_2)$ $(x \geq 0)$ is given by*

$$\int_0^{+\infty} e^{-ix\xi} A^\theta(\Sigma_1, \Sigma_2)dx = \frac{1}{2}\big[DM^\theta(\Sigma_1) + TM^\theta(\Sigma_2)\big](\xi) \qquad (3.4)$$

where DM^θ, TM^θ are the integral operators defined by

$$\begin{cases} DM^\theta(\hat{f})(\xi) = \displaystyle\int_{\Gamma_0} DM^\theta(\xi,\zeta)\hat{f}(\zeta)d\zeta \\[2mm] TM^\theta(\hat{f})(\xi) = \displaystyle\int_{\Gamma_0} TM^\theta(\xi,\zeta)\hat{f}(\zeta)d\zeta \end{cases} , \quad f \in \mathcal{A}^3. \qquad (3.5)$$

The kernels $DM^\theta(\xi,\zeta)$, $TM^\theta(\xi,\zeta)$ are given by

$$DM^\theta(\xi,\zeta) = \frac{1}{2i\pi}\,\frac{1}{\xi - \zeta}dm(\zeta e^{i\theta}) \qquad (3.6)$$

with

$$dm(z) = \begin{bmatrix} -1 & A & 0 \\ B & -1 & C \\ 0 & D & -1 \end{bmatrix} \qquad (3.7)$$

$$\begin{cases} A(z) = \dfrac{z}{\zeta_T(z)}(1 - 2\mu Q(z)); \quad B(z) = -\dfrac{z}{\zeta_L(z)}(1 - 2\mu Q(z)) \\[3mm] C(z) = \dfrac{\rho}{\zeta_0}(z); \quad D(z) = -\dfrac{Q(z)}{\zeta_T(z)} \\[3mm] Q(z) = \zeta_L(z)\zeta_T(z) + z^2 \end{cases} \qquad (3.8)$$

$$TM^\theta(\xi,\zeta) = \frac{1}{2i\pi}\sum_* D_*^\theta(\xi,\zeta)tm_*(\zeta e^{i\theta}), \quad * \in \{0, L, T\} \qquad (3.9)$$

with

$$D_*^\theta(\xi,\zeta) = \frac{1}{\xi - [\cos\varphi\zeta + \sin\varphi\zeta_*^\theta(\zeta)]}.$$

The three matrices $tm_L(z)$, $tm_T(z)$, $tm_0(z)$ are of rank 1 and are given by

$$\begin{cases} tm_L(z) = \left(-\dfrac{\cos\chi}{\sin\chi}\mathbf{f}_L, \mathbf{f}_L, 0\right); \quad \mathbf{f}_L = \begin{bmatrix} -\mu\sin 2\psi \\ \mu - 1 + \mu\cos 2\psi \\ -\sin\psi \end{bmatrix} \\[4mm] z = \nu_L\cos\chi, \quad \zeta_L(z) = -\nu_L\sin\chi; \quad \psi = \varphi + \chi \end{cases} \qquad (3.10)$$

$$\begin{cases} tm_T(z) = \left(-f_T; -f_T\dfrac{\cos\chi}{\sin\chi}; 0\right); \quad \mathbf{f}_T = \begin{bmatrix} \cos 2\psi \\ \sin 2\psi \\ \nu_T\cos\psi \end{bmatrix} \\[4mm] z = \nu_T\cos\chi, \quad \zeta_T(z) = -\nu_T\sin\chi; \quad \psi = \varphi + \chi \end{cases} \tag{3.11}$$

$$\begin{cases} tm_0(z) = (0,0,-f_0); \quad \mathbf{f}_0 = \begin{bmatrix} 0 \\ \dfrac{\rho}{\nu_0\sin\chi} \\ \dfrac{\sin\psi}{\sin\chi} \end{bmatrix} \\[4mm] z = \nu_0\cos\chi, \quad \zeta_0(z) = -\nu_0\sin\chi; \quad \psi = \varphi + \chi. \end{cases} \tag{3.12}$$

Proof. We decompose in two parts the boundary operator $A^\theta(\Sigma_1, \Sigma_2)$

$$A^\theta(\Sigma_1, \Sigma_2) = A_1^\theta(\Sigma_1) + A_2^\theta(\Sigma_2)$$

where $A_j^\theta(\Sigma_j)$ depends only on (v_j^θ, h_j^θ), $j = 1, 2$. We have

$$A_j^\theta(\Sigma_j) = \begin{cases} (\lambda \operatorname{div} v_j^\theta \operatorname{Id} + 2\mu\varepsilon(v_j^\theta)) \cdot \mathbf{n} - ie^{-i\theta}\rho h_j^\theta \mathbf{n}|_{\Gamma_1} \quad (\in \mathbb{C}^2) \\ ie^{-i\theta}v_j^\theta \cdot \mathbf{n} - \operatorname{grad} h_j^\theta \cdot \mathbf{n}|_{\Gamma_1} \quad (\in \mathbb{C}). \end{cases} \tag{3.13}$$

Expressing the r.h.s of (3.13) with the partial derivatives of $v_1^\theta = (v_1^{\theta,x}, v_1^{\theta,y})$, h_1^θ, evaluated at the point $(x,0) \in \Gamma_1$, yields

$$A_1^\theta(\Sigma_1) = \begin{cases} \mu\left(\dfrac{\partial v_1^{\theta,x}}{\partial y} + \dfrac{\partial v_1^{\theta,y}}{\partial x}\right) \quad \text{(component on } \mathbf{t}) \\[3mm] (\lambda + 2\mu)\dfrac{\partial v_1^{\theta,y}}{\partial y} + \lambda\dfrac{\partial v_1^{\theta,x}}{\partial x} - ie^{-i\theta}\rho h_1 \quad \text{(component on } \mathbf{n}). \\[3mm] ie^{-i\theta}v_1^{\theta,y} - \dfrac{\partial h_1^\theta}{\partial y} \end{cases} \tag{3.14}$$

On the other hand, the functions $v_2^\theta = (v_2^{\theta,x'}, v_2^{\theta,y'})$ and h_2^θ depend on the variables (x', y') (cf. Fig. 1.1). Using the change of variables

$$x' = x\cos\varphi + y\sin\varphi, \quad y' = x\sin\varphi - y\cos\varphi. \tag{3.15}$$

we get for the three components of $A_2^\theta(\Sigma_2)$:

$$\begin{cases} 2\mu\sin\varphi\cos\varphi\left(\dfrac{\partial v_2^{\theta,x'}}{\partial x'} - \dfrac{\partial v_2^{\theta,y'}}{\partial y'}\right) + \mu(\sin^2\varphi - \cos^2\varphi)\left(\dfrac{\partial v_2^{\theta,x'}}{\partial y'} + \dfrac{\partial v_2^{\theta,y'}}{\partial x'}\right) \\[3mm] (\lambda + 2\mu\sin^2\varphi)\dfrac{\partial v_2^{\theta,x'}}{\partial x'} + (\lambda + 2\mu\cos^2\varphi)\dfrac{\partial v_2^{\theta,y'}}{\partial y'} \\[3mm] \qquad\qquad - 2\mu\cos\varphi\sin\varphi\left(\dfrac{\partial v_2^{\theta,x'}}{\partial y'} + \dfrac{\partial v_2^{\theta,y'}}{\partial x'}\right) - ie^{-i\theta}\rho h_2^\theta(\mathbf{n}) \\[3mm] ie^{-i\theta}\left(\sin\varphi v_2^{\theta,x'} - \cos\varphi v_2^{\theta,y'}\right) - \left(\sin\varphi\dfrac{\partial h_2^\theta}{\partial x'} - \cos\varphi\dfrac{\partial h_2^\theta}{\partial y'}\right) \end{cases} \tag{3.16}$$

where all the functions are evaluated at the point $(x \cos \varphi, x \sin \varphi)$, $x \geq 0$.

We first prove that $\mathcal{F}(A_1^\theta)(\xi) = \frac{1}{2} D M^\theta(\Sigma_1)(\xi)$, where $\mathcal{F}$ stands for the Fourier transform. Differentiating (2.57-2.58) with respect to x, y at the point $(x, 0)$, $x \geq 0$, yields

$$\frac{\partial v_1^\theta}{\partial x}(x, 0^+) = -\frac{1}{4\pi} \int_{\Gamma_0} e^{ix\xi} \begin{bmatrix} \hat{\alpha}_1 \left[\frac{\xi^2 e^{2i\theta}}{\zeta_L(\xi e^{i\theta})} + \zeta_T(\xi e^{i\theta}) \right] \xi e^{i\theta} \\ \hat{\beta}_1 \left[\frac{\xi^2 e^{2i\theta}}{\zeta_T(\xi e^{i\theta})} + \zeta_L(\xi e^{i\theta}) \right] \xi e^{i\theta} \end{bmatrix} d\xi \quad (3.17)$$

$$\frac{\partial v_1^\theta}{\partial y}(x, 0^+) = -\frac{1}{4\pi} \int_{\Gamma_0} e^{ix\xi} \begin{bmatrix} \hat{\alpha}_1 \nu_T^2 + \hat{\beta}_1 (\zeta_L(\xi e^{i\theta}) - \zeta_T(\xi e^{i\theta})) \xi e^{i\theta} \\ \hat{\alpha}_1 (\zeta_L(\xi e^{i\theta}) - \zeta_T(\xi e^{i\theta})) \xi e^{i\theta} + \hat{\beta}_1 \nu_L^2 \end{bmatrix} d\xi \quad (3.18)$$

$$\frac{\partial h_1^\theta}{\partial y}(x, 0^+) = \frac{1}{4\pi} \int_{\Gamma_0} e^{ix\xi} \hat{\gamma}_1(\xi) d\xi. \quad (3.19)$$

According to Lemma 2.2, each of these traces on Γ_1 is in the space L^2_{loc} $(x \geq 0)$ and is a tempered distribution near $+\infty$. It is easy to check that replacing in (3.14) each partial derivative by its value given by (3.17-3.19) allows to rewrite $A_1^\theta(\Sigma_1)$ as

$$A_1^\theta(\Sigma_1)(x \geq 0) = \frac{1}{4\pi} \int_{\Gamma_0} e^{ix\xi} dm(\xi e^{i\theta}) \cdot \Sigma_1(\xi) d\xi \quad (\in \mathbb{C}^3) \quad (3.20)$$

where the matrix $dm(z)$ is given by (3.7). Finally, we take the Fourier transform of $A_1^\theta(\Sigma_1)(x \geq 0)$. Using the identity

$$\int_0^{+\infty} e^{-ix(\xi - \zeta)} dx = \frac{1}{i(\xi - \zeta)}, \quad \text{Im}\, \xi < 0, \quad \text{Im}\, \zeta > 0. \quad (3.21)$$

We deduce from (3.20) that

$$\int_0^{+\infty} e^{-ix\xi} A_1^\theta(x) dx = \frac{1}{4i\pi} \int_{\Gamma_0} \frac{1}{(\xi - \zeta)} dm(\zeta e^{i\theta}) \cdot \Sigma_1(\zeta) d\zeta \quad (3.22)$$

$$= \frac{1}{2} D M^\theta(\Sigma_1)(\xi).$$

The computations to prove that $\mathcal{F}(A_2^\theta)(\xi) = \frac{1}{2} T M^\theta(\Sigma_2)(\xi)$ are a little more tedious, but follow the same scheme. The substitution of the partial derivatives $\frac{\partial v_2^\theta}{\partial x'}, \frac{\partial v_2^\theta}{\partial y'}, \frac{\partial h_2^\theta}{\partial x'}, \frac{\partial h_2^\theta}{\partial y'}$ by their values obtained by differentiating (2.59-2.60) gives

$$A_2^\theta(\Sigma_2)(x \geq 0) = \frac{1}{4\pi} \int_{\Gamma_0} \left[e^{ix(\cos \varphi \xi + \sin \varphi \zeta_L^\theta)} tm_L(\xi e^{i\theta}) \right.$$
$$\left. + e^{ix(\cos \varphi \xi + \sin \varphi \zeta_T^\theta)} tm_T(\xi) + e^{ix(\cos \varphi \xi + \sin \varphi \zeta_0^\theta)} tm_0(\xi e^{i\theta}) \right] \cdot \Sigma_2(\xi) d\xi \quad (3.23)$$

The 3×3 matrices $tm_L(\zeta)$, $tm_T(\zeta)$, $tm_0(\zeta)$ are of rank 1, and are given by (we note $\zeta_* = \zeta_*(\zeta)$ for $* = L, T, 0$)

$$tm_L(\zeta) = \left[\frac{\zeta}{\zeta_L} \mathbf{f}_L, \mathbf{f}_L, 0 \right], \quad \mathbf{f}_L = \begin{bmatrix} \mu[\cos 2\varphi(2\zeta\zeta_L) - \sin 2\varphi(\zeta^2 - \zeta_L^2)] \\ (\mu - 1) + \mu[2\sin 2\varphi\zeta\zeta_L + \cos 2\varphi(\zeta^2 - \zeta_L^2)] \\ -\sin\varphi\zeta + \cos\varphi\zeta_L \end{bmatrix}$$

$$(3.24)$$

$$tm_T(\zeta) = \left[-\mathbf{f}_T; \frac{\zeta}{\zeta_T} \mathbf{f}_T; 0 \right], \quad \mathbf{f}_T = \begin{bmatrix} \mu[2\sin 2\varphi\zeta\zeta_T + \cos 2\varphi(\zeta^2 - \zeta_T^2)] \\ \mu[-2\cos 2\varphi\zeta\zeta_T + \sin 2\varphi(\zeta^2 - \zeta_T^2)] \\ \cos\varphi\zeta + \sin\varphi\zeta_T \end{bmatrix}$$

$$(3.25)$$

$$tm_0(\zeta) = [0; 0; -\mathbf{f}_0], \quad \mathbf{f}_0 = \begin{bmatrix} 0 \\ -\frac{\rho}{\zeta_0} \\ \cos\varphi - \sin\varphi\frac{\zeta}{\zeta_0} \end{bmatrix}.$$

$$(3.26)$$

We take again the Fourier transform of $A_2^\theta(x)$ along the face 1. Using the identity

$$\int_0^{+\infty} e^{-ix(\xi - (\cos\varphi\zeta + \sin\varphi\zeta_*^\theta))} = \frac{1}{i(\xi - (\cos\varphi\zeta + \sin\varphi\zeta_*^\theta))}, \quad \text{Im } \xi < 0, \ \text{Im } \zeta > 0$$

$$(3.27)$$

and defining the function $D_*^\theta(\xi, \zeta)$ by

$$D_*^\theta(\xi, \zeta) = \frac{1}{\xi - (\cos\varphi\zeta + \sin\varphi\zeta_*^\theta)}, \quad * \in \{L, T, 0\}, \ \theta \in [0, \pi[.$$

$$(3.28)$$

we get

$$\int_0^{+\infty} e^{-ix\xi} A_2^\theta(x) dx = \frac{1}{4i\pi} \int_{\Gamma_0} TM^\theta(\xi, \zeta) \cdot \Sigma_2(\zeta) d\zeta$$

$$(3.29)$$

$$= \frac{1}{2} TM^\theta(\Sigma_2)(\xi)$$

where $TM^\theta(\xi, \zeta)$ is defined by

$$TM^\theta(\xi, \zeta) = \sum_{*\in\{L,T,0\}} D_*^\theta(\xi, \zeta) tm_*(\zeta e^{i\theta}).$$

$$(3.30)$$

Finally, the formulas (3.10-3.12) are easily deduced from the substitution in (3.24-3.26) of $(\zeta, \zeta_*(\zeta))$ by $(\nu_* \cos\chi, -\nu_* \sin\chi)$ and by using the classical trigonometry formulas. ∎

It is now easy to derive the following result. Recall that the case $\theta = 0$ corresponds to the outgoing solutions of (2.12-2.15).

Lemma 3.2. *For* $\theta \in [0, \pi[$, *the functions* (v^θ, h^θ) *are a solution of the system* (S_θ) *if and only if the spectral function* $\Sigma = (\Sigma_1, \Sigma_2)$ *is solution of the system*

$$
S_\theta \begin{cases} DM^\theta \cdot \Sigma_1 + TM^\theta \cdot \Sigma_2 = \dfrac{W_1^\theta}{\xi - \nu_0 \cos\theta_{in}} \\[2mm] TM^\theta \cdot \Sigma_1 + DM^\theta \cdot \Sigma_2 = \dfrac{W_2^\theta}{\xi - \nu_0 \cos(\theta_{in} + \varphi)} \end{cases} \qquad \forall \xi, \ \operatorname{Im}\xi < 0 \quad (3.31)
$$

where $W_1^\theta, W_2^\theta \in \mathbb{C}^3$ *are*

$$
W_1^\theta = \begin{bmatrix} 0 \\ \rho e^{-i\theta} \\ -\nu_0 \sin\theta_{in} \end{bmatrix} ; \quad W_2^\theta = \begin{bmatrix} 0 \\ \rho e^{-i\theta} \\ \nu_0 \sin(\theta_{in} + \varphi) \end{bmatrix} . \tag{3.32}
$$

Proof. The right hand-side in (2.21) is the vector $R(x, y)$ given by

$$
R(x, y) = \begin{bmatrix} 0 \\ \frac{i\rho e^{-i\theta}}{2} e^{i\nu_0(x\cos\theta_{in} - y\sin\theta_{in})} \\ \frac{i\nu_0}{2}(\cos\theta_{in} n_x - \sin\theta_{in} n_y)e^{i\nu_0(x\cos\theta_{in} - y\sin\theta_{in})} \end{bmatrix}, \quad (x, y) \in \Gamma \tag{3.33}
$$

Along the face 1 of the wedge, it reduces to

$$
(x, 0) \in \Gamma_1 \xrightarrow{R_1} \begin{bmatrix} 0 \\ \frac{i\rho e^{-i\theta}}{2} e^{i\nu_0 x \cos\theta_{in}} \\ -\frac{i\nu_0 \sin\theta_{in}}{2} e^{i\nu_0 x \cos\theta_{in}} \end{bmatrix} . \tag{3.34}
$$

Its Fourier transform is

$$
\int_0^{+\infty} e^{-ix\xi} R_1(x) dx = \frac{1}{2} \frac{W_1^\theta}{\xi - \nu_0 \cos\theta_{in}} \tag{3.35}
$$

where W_1^θ is given by (3.32).

We deduce now from (3.4) and (3.35) that

$$
DM^\theta(\Sigma_1) + TM^\theta(\Sigma_2) = \frac{W_1^\theta}{\xi - \nu_0 \cos\theta_{in}}. \tag{3.36}
$$

The second equation in (3.31) is obtained in the same way, by remarking that each computation for the face 1 is transformed in the corresponding computation for the face 2, by replacing $(x, y, \mathbf{t}, \mathbf{n}, \theta_{in})$ by $(x', y', \mathbf{t}', \mathbf{n}', -(\theta_{in} + \varphi))$ and the subscripts $(1, 2)$ by $(2, 1)$. ■

In the sequel, we call $(S) = (S_{\theta=0})$ the system in the spectral function $\Sigma = (\Sigma_1, \Sigma_2)$, associated to an outgoing solution of (2.12-2.15)

$$(S) \begin{cases} DM \cdot \Sigma_1 + TM \cdot \Sigma_2 = \dfrac{W_1}{\xi - \nu_0 \cos \theta_{in}} \\[4mm] TM \cdot \Sigma_1 + DM \cdot \Sigma_2 = \dfrac{W_2}{\xi - \nu_0 \cos(\theta_{in} + \varphi)} \end{cases} \tag{3.37}$$

where

$$DM = DM^{\theta=0}, \quad TM = TM^{\theta=0}; \qquad W_j = W_j^{\theta=0}, \quad j = 1, 2. \tag{3.38}$$

3.2. The Equations Solid/Fluid Coupled by a Plane Surface

The aim of this section is to give some properties of the 3×3 matrix $dm(z)$ occuring in the kernel $DM(\xi, \zeta)$ given by (3.6). This matrix is in fact related to the problem (2.12-2.15), when the interface between the solid and the fluid is a plane interface instead of a wedge (cf. Fig. 3.1).

As in §. 2.3 we introduce, for $\theta \in [0, \pi[$ the system indexed by a parameter $\theta \in [0, \pi[$

$$\begin{cases} (E + e^{-2i\theta})v^\theta = 0 & \text{in } y > 0 \quad (\Omega_s) \\ (\Delta + \nu_0^2 e^{-2i\theta})h^\theta = 0 & \text{in } y < 0 \quad (\Omega_f) \\ (\lambda \operatorname{div} v^\theta + 2\mu\varepsilon(v^\theta))\mathbf{n} - i e^{-i\theta}\rho h^\theta \mathbf{n} = 0 & \text{on } y = 0 \quad (\Gamma) \\ (i e^{-i\theta}v^\theta \cdot \mathbf{n} - \operatorname{grad} v^\theta \cdot \mathbf{n}) = 0 & \text{on } y = 0 \quad (\Gamma) \end{cases} \tag{3.39}_\theta$$

We call *interface-wave* a solution of $(3.39)_\theta$ of the form

$$v^\theta(x, y) = e^{ix\xi}v_\xi^\theta(y), \ y > 0 \ ; \ h^\theta(x, y) = e^{ix\xi}h_\xi^\theta(y), \ y < 0 \tag{3.40}$$

where $\xi \in \mathbb{R}$, and v_ξ^θ, (resp. h_ξ^θ) is in the space $S(\mathbb{R}^+)$, of the functions in $C^\infty([0, +\infty[)$ which are rapidly decreasing with their derivatives at $+\infty$, (resp. the functions in $C^\infty(]-\infty, 0])$ which are rapidly decreasing with their derivatives at $-\infty$).

The first result is

Lemma 3.3. *For $\theta \in]0, \pi[$, the problem $(3.39)_\theta$ has no non trivial solutions of interface-wave type.*

Proof. The Green formula for the elasticity operator E in the half-plane $y > 0$ reads, for any $\mathbb{C}^2$-valued sufficiently regular functions u, v :

$$\iint_{y>0} -Ev \cdot \overline{u} = \iint_{y>0} \lambda \operatorname{div} v \cdot \operatorname{div} \overline{u} + 2\mu \operatorname{tr}(\varepsilon(v)\varepsilon^*(u))$$
$$+ \int_{y=0} \left[(\lambda \operatorname{div} v \operatorname{Id} + 2\mu\varepsilon(\sigma)) \cdot \mathbf{n} \right] \cdot \overline{u} \tag{3.41}$$

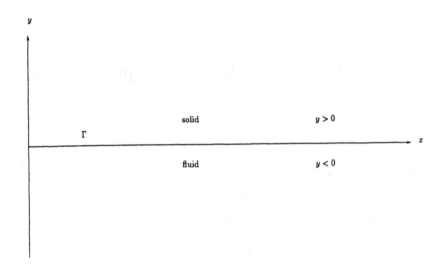

Fig. 3.1 The coupled problem by a plane interface

For the laplacian operator, we have, for any $\mathbb{C}$−valued functions g, h :

$$\iint_{y<0} -\Delta h \overline{g} = \iint_{y<0} \nabla h \cdot \nabla \overline{g} - \int_{y=0} \frac{\partial h}{\partial n} \overline{g}. \tag{3.42}$$

Let (v, h) be of interface-wave type

$$v(x,y) = e^{ix\xi} v_\xi(y), \quad h(x,y) = e^{ix\xi} h_\xi(y), \quad \xi \in \mathbb{R}. \tag{3.43}$$

For any $P(\partial_x, \partial_y)$ differential operator with constant coefficients, we call $P_\xi(\partial_y) = P(i\xi, \partial_x)$ the operator defined by

$$P(e^{ix\xi} f(y)) = e^{ix\xi} (P_\xi f)(y) \tag{3.44}$$

by taking $u(x,y) = \varphi(x)v_\xi(y)$, $g(x,y) = \varphi(x)h_\xi(y)$, where φ is any C^∞ function with compact support, we deduce easily that

$$\int_{y>0} -E_\xi v_\xi \cdot \overline{v}_\xi = \int_{y>0} \left[\lambda \operatorname{div}_\xi v_\xi \cdot \operatorname{div}_\xi \overline{v}_\xi + 2\mu \operatorname{tr}(\varepsilon_\xi(v_\xi)\varepsilon_\xi^*(v_\xi)) \right] \tag{3.45}$$
$$+ \left(\lambda \operatorname{div} v_\xi(0) \operatorname{Id} + 2\mu\varepsilon_\xi(v_\xi(0)) \right) \cdot \mathbf{n} \cdot v_\xi(0)$$

$$\int_{y<0} -\Delta_\xi h_\xi \cdot \overline{h}_\xi = \int_{y<0} |\nabla_\xi h_\xi|^2 - (\operatorname{grad}_\xi h_\xi(0) \cdot \mathbf{n})\overline{h}_\xi(0). \tag{3.46}$$

We define the quantities, $e_1, e_2 \in \mathbb{R}^+$ by

$$e_1 = \int_{y>0} \left(\lambda |\operatorname{div}_\xi v_\xi|^2 + 2\mu \operatorname{tr}(\varepsilon_\xi(v_\xi)\varepsilon_\xi^*(v_\xi)) \right) + \rho \int_{y<0} |\nabla_\xi h_\xi|^2 \tag{3.47}$$

$$e_2 = \int_{y>0} |v_\xi|^2 + \rho\nu_0^2 \int_{y<0} |h_\xi|^2. \tag{3.48}$$

If (v, h) is solution of (3.39), then we have

$$e^{i\theta} e_1 - e^{-i\theta} e_2 = i\rho\big(\overline{h}_\xi(0)v_\xi(0) \cdot \mathbf{n} - h_\xi(0)\overline{v}_\xi(0) \cdot \mathbf{n}\big). \tag{3.49}$$

Taking the imaginary part of each side yields $\sin\theta(e_1 + e_2) = 0$. Thus $e_1 = e_2 = 0$, which ensures $u_\xi = 0$, $g_\xi = 0$. ∎

Lemma 3.4 (Scholte-Stoneley interface-wave).

(i) *The problem* $(3.39)_\theta$ *does have interface-wave solutions only for* $\theta = 0$ *and* $|\xi| > \nu_0$. *They are of the form*

$$v(x,y) = e^{ix\xi} v_\xi(y); \quad h(x,y) = e^{ix\xi} h_\xi(y)$$

$$\begin{cases} v_\xi(y) = \dfrac{i}{2}\left\{ e^{iy\zeta_L(\xi)}\left(\hat{\beta}_\xi + \dfrac{\xi}{\zeta_L(\xi)}\hat{\alpha}_\xi \right) \begin{bmatrix} \xi \\ \zeta_L(\xi) \end{bmatrix} \right. \\ \qquad\qquad \left. + e^{iy\zeta_T(\xi)}\left(-\hat{\alpha}_\xi + \dfrac{\xi}{\zeta_T(\xi)}\hat{\beta}_\xi \right) \begin{bmatrix} -\zeta_T(\xi) \\ \xi \end{bmatrix} \right\}, \quad y > 0 \\ h_\xi(y) = \dfrac{i}{2}\left\{ e^{-iy\zeta_0(\xi)} \dfrac{\hat{\gamma}_\xi}{\zeta_0(\xi)} \right\}, \quad y < 0 \end{cases}$$

$$\tag{3.50}$$

where $(\hat{\alpha}_\xi, \hat{\beta}_\xi, \hat{\gamma}_\xi)^T \in \mathbb{C}^3$ *is a non null solution of*

$$dm(e^{i\theta}\xi) \begin{pmatrix} \hat{\alpha}_\xi \\ \hat{\beta}_\xi \\ \hat{\gamma}_\xi \end{pmatrix} = 0, \qquad \xi \in \mathbb{R}. \tag{3.51}$$

(ii) *The characteristic equation*

$$dm(\xi) = 0, \qquad \xi \in \mathbb{R} \tag{3.52}$$

does possess only two solutions $\pm\nu_S$. The velocity c_S defined by $\nu_S = \frac{c_L}{c_S}$, is the Scholte-Stoneley velocity.

An immediate consequence of this lemma is that the only interface-wave solutions are the Scholte-Stoneley waves, that is the waves (3.50) with $\xi = \pm\nu_S$ and $(\hat{\alpha}_S, \hat{\beta}_S, \hat{\gamma}_S)^T \in \mathbb{C}^3 \setminus \{0\}$ any non null vector in the null-space of $dm(\pm\nu_S)$.

Proof.

(i) Suppose $v^\theta(x, y) = e^{ix\xi}v_\xi^\theta(y)$, $y > 0$, $h^\theta(x, y) = e^{ix\xi}h_\xi^\theta(y)$, $y < 0$ is an interface-wave solution of (3.39). We call still v_ξ^θ (resp. h_ξ^θ) the prolongation by symmetry of v_ξ^θ to $y < 0$, (resp. of h_ξ^θ to $y > 0$). If $P(x, y)$ is a linear differential operator in (x, y), we call $P_\xi(y)$ the operator defined by

$$P(e^{ix\xi}f(y)) = e^{ix\xi}P_\xi f(y).$$

The functions v_ξ^θ, h_ξ^θ are thus solutions of

$$(E_\xi + e^{-2i\theta}\operatorname{Id})v_\xi^\theta = 0, \quad (\Delta_\xi + \nu_0^2 e^{-2i\theta})h_\xi^\theta = 0, \quad |y| > 0. \tag{3.53}$$

We deduce from the continuity of v_ξ^θ, h_ξ^θ in $y = 0$ that there exist $\hat{\alpha}_\xi$, $\hat{\beta}_\xi$, $\hat{\gamma}_\xi \in \mathbb{C}$ such that

$$\begin{cases} (E_\xi + e^{-2i\theta}\operatorname{Id})v_\xi^\theta = -\begin{pmatrix} \hat{\alpha}_\xi \\ \hat{\beta}_\xi \end{pmatrix} \otimes \delta(y = 0) \\ (\Delta_\xi + \nu_0^2 e^{-2i\theta})h_\xi^\theta = -\hat{\gamma}_\xi \otimes \delta(y = 0) \end{cases} \tag{3.54}$$

Hence the Fourier transform of v_ξ^θ, h_ξ^θ are (cf. (2.44), (2.53))

$$\hat{v}_\xi^\theta(\eta) = \left(M(\xi, \eta) - e^{-2i\theta}\right)^{-1}\begin{pmatrix} \hat{\alpha}_\xi \\ \hat{\beta}_\xi \end{pmatrix} \tag{3.55}$$

$$= \frac{(\xi\hat{\alpha}_\xi + \eta\hat{\beta}_\xi)\nu_L^2}{(\xi^2 + \eta^2)(\xi^2 + \eta^2 - \nu_L^2 e^{-2i\theta})}\begin{pmatrix} \xi \\ \eta \end{pmatrix}$$

$$+ \frac{(-\eta\hat{\alpha}_\xi + \xi\hat{\beta}_\xi)\nu_T^2}{(\xi^2 + \eta^2)(\xi^2 + \eta^2 - \nu_T^2 e^{-2i\theta})}\begin{pmatrix} -\eta \\ \xi \end{pmatrix}$$

$$\hat{h}_\xi^\theta(\eta) = \frac{\hat{\gamma}_\xi}{(\xi^2 + \eta^2 - \nu_0^2 e^{-2i\theta})}. \tag{3.56}$$

• **1st case** : $\theta \in]0, \pi[$ or ($\theta = 0$ and $|\xi| > \nu_0$). In this case, $\hat{v}_\xi^\theta, \hat{h}_\xi^\theta$ are integrable, continuous functions of the variable $\eta \in \mathbb{R}$. The inverse Fourier transform is, for $y \in \mathbb{R}$

$$\begin{cases} v_\xi^\theta(y) = \dfrac{1}{2\pi} \displaystyle\int_\mathbb{R} e^{iy\eta} (M(\xi,\eta) - e^{-2i\theta})^{-1} \begin{pmatrix} \hat{\alpha}_\xi \\ \hat{\beta}_\xi \end{pmatrix} d\eta \\[4mm] h_\xi^\theta(y) = \dfrac{1}{2\pi} \displaystyle\int_\mathbb{R} e^{iy\eta} \dfrac{\hat{\gamma}_\xi}{(\eta^2 + \xi^2) - \nu_0^2 e^{-2i\theta}} d\eta. \end{cases} \tag{3.57}$$

Recalling that $\xi^2 - \nu_*^2 e^{-2i\theta} = \zeta_*^\theta(\xi)^2$, we get by the Cauchy formula

$$\begin{cases} v_\xi^\theta(y) = \dfrac{i}{2} e^{2i\theta} \left[e^{iy\zeta_L^\theta(\xi)} \left(\hat{\alpha}_\xi \dfrac{\xi}{\zeta_L^\theta(\xi)} + \hat{\beta}_\xi \right) \begin{bmatrix} \xi \\ \zeta_L^\theta(\xi) \end{bmatrix} \right. \\[4mm] \qquad\qquad \left. + e^{iy\zeta_T^\theta(\xi)} \left(-\hat{\alpha}_\xi + \dfrac{\xi}{\zeta_T^\theta(\xi)} \hat{\beta}_\xi \right) \begin{bmatrix} -\zeta_T^\theta(\xi) \\ \xi \end{bmatrix} \right] \\[4mm] h_\xi^\theta(y) = \dfrac{i}{2} e^{-i\zeta_0^\theta(\xi)y} \dfrac{\hat{\gamma}_\xi}{\zeta_0^\theta(\xi)}. \end{cases} \tag{3.58}$$

We check easily that $v_\xi^\theta \in \mathcal{S}(\mathbb{R}^+)$, $h_\xi^\theta \in \mathcal{S}(\mathbb{R}^-)$.

- **2nd case :** $\theta = 0$ and $|\xi| \le \nu_0$.

In this case, we have that $\hat{h}_\xi(\eta) = \frac{\hat{\gamma}_\xi}{\eta^2 - \zeta^2 - \zeta_0^2(\xi)}$. Therefore

$$h_\xi(y) = \begin{cases} -\dfrac{\hat{\gamma}_\xi}{2\zeta_0(\xi)} \sin(|y|\zeta_0(\xi)) & \text{if } |\xi| < \nu_0 \\[4mm] \dfrac{i\hat{\gamma}_\xi}{2} |y| & \text{if } |\xi| = \nu_0. \end{cases} \tag{3.59}$$

In all cases, $h_\xi \notin \mathcal{S}(\mathbb{R}^-)$, hence no interface-wave equation does exist in this case.

Finally, it remains to translate the two boundary equations in $(3.39)_{\theta=0}$ in terms of v_ξ^θ, h_ξ^θ. They can be written as

$$\frac{1}{2} \begin{bmatrix} -1 & \frac{\xi}{\zeta_T^\theta} - 2\mu e^{2i\theta} \xi (\zeta_L^\theta + \frac{\xi^2}{\zeta_T^\theta}) & 0 \\[3mm] -\frac{\xi}{\zeta_L^\theta} - 2\mu e^{2i\theta} \xi (\zeta_T^\theta + \frac{\xi^2}{\zeta_L^\theta}) & -1 & \rho \frac{e^{-i\theta}}{\zeta_0^\theta} \\[3mm] 0 & -e^{i\theta} (\zeta_L^\theta + \frac{\xi^2}{\zeta_T^\theta}) & -1 \end{bmatrix} \begin{bmatrix} \hat{\alpha}_\xi \\ \hat{\beta}_\xi \\ \hat{\gamma}_\xi \end{bmatrix} = 0 \tag{3.60}$$

or equivalently

$$dm(e^{i\theta}\xi) \begin{bmatrix} \hat{\alpha}_\xi \\ \hat{\beta}_\xi \\ \hat{\gamma}_\xi \end{bmatrix} = 0. \tag{3.61}$$

The conclusion of the part (i) of Lemma 3.4 follows now from Lemma 3.3.

(ii) The matrix function $z \mapsto dm(z) \in \mathbb{M}_3(\mathbb{C})$, defined by (3.7), (3.8), is an holomorphic function in the domain

$$V = \mathbb{C} \setminus \{z \in \mathbb{R}, |z| \geq \nu_L = 1\}. \tag{3.62}$$

Its determinant is

$$\det(dm(z)) = \Delta(z) = -\frac{Q}{\zeta_L \zeta_T}\left[1 - 4\mu z^2 + 4\mu^2 z^2 Q + \rho\frac{\zeta_L}{\zeta_0}\right] \tag{3.63}$$

and the inverse matrix is

$$dm(z)^{-1} = \frac{1}{\Delta}\begin{bmatrix} 1 - CD & A & AC \\ B & 1 & C \\ DB & D & 1 - AB \end{bmatrix}. \tag{3.64}$$

The matrix $dm(z)$ is well defined for z such that $\zeta_*(z) \neq 0$, $* \in \{L, T, 0\}$, or equivalently for $z \notin \{\pm 1 = \pm\nu_L, \pm\nu_T, \pm\nu_0\}$. The functions $\zeta_{L,T,0}(\xi)$ being even functions in $\xi \in \mathbb{R}$, $\Delta(\xi)$ is also a even function. We check easily that $Q(\xi) \neq 0$ for $\xi \in \mathbb{R}$, hence the characteristic equation $\Delta(\xi) = 0$ is equivalent to the equation

$$S(\xi) \overset{\text{def}}{=} 1 - 4\mu\xi^2 + 4\mu^2\xi^2 Q(\xi) + \rho\frac{\zeta_L}{\zeta_0} = 0. \tag{3.65}$$

(3.65) is the *Scholte-Stoneley equation*. Note that $S(\xi)$ is also denoted $\delta(\xi)$ elsewhere. $S(\xi)$ can be rewritten as

$$S(\xi) = (1 - 2\mu\xi^2)^2 + 4\mu^2\xi^2\zeta_L\zeta_T + \rho\frac{\zeta_L}{\zeta_0}. \tag{3.66}$$

Therefore, we have the four cases

- $0 \leq \xi \leq 1$

  We have $S(\xi) > 0$ in this case.

- $1 \leq \xi \leq \nu_T$

  We have

$$\text{Re}\, S(\xi) = (1 - 2\mu\xi^2)^2 \;;\; \text{Im}\, S(\xi) = -\sqrt{\xi^2 - 1}\left(4\mu^2\xi^2\sqrt{\nu_T^2 - \xi^2} + \frac{\rho}{\sqrt{\nu_0^2 - \xi^2}}\right) \tag{3.67}$$

Therefore $S(\xi) \neq 0$, because $\mu = \frac{\mu}{\lambda + 2\mu} \in \,]0, \frac{1}{2}[$.

- $\nu_T \leq \xi \leq \nu_0$

  We have

$$\text{Im}\, S(\xi) = -\rho\frac{\sqrt{\xi^2 - 1}}{\sqrt{\nu_0^2 - \xi^2}} < 0. \tag{3.68}$$

- $\xi > \nu_0$

  We have

$$S(\xi) = (1 - 2\mu\xi^2)^2 - 4\mu^2\xi^2\sqrt{\xi^2 - 1}\sqrt{\xi^2 - \nu_T^2} + \rho\frac{\sqrt{\xi^2 - 1}}{\sqrt{\xi^2 - \nu_0^2}} = s(\xi^2). \quad (3.69)$$

Defining $R(\xi)$ as

$$R(\xi) = 1 - 4\mu\xi^2 + 4\mu^2\xi^2 Q(\xi) = (1 - 2\mu\xi^2)^2 + 4\mu^2\xi^2\zeta_L\zeta_T \quad (3.70)$$

then $R(\xi) = 0$ is the *Rayleigh equation*. As previously, $R(\xi) \neq 0$ for $0 \leq \xi \leq \nu_T$ and

$$\xi \geq \nu_T \Longrightarrow R(\xi) = (1 - 2\mu\xi^2)^2 - 4\mu^2\xi^2\sqrt{\xi^2 - 1}\sqrt{\xi^2 - \nu_T^2} = r(\xi^2). \quad (3.71)$$

By setting $x = \nu_T^2 z$, we check that

$$R(x) = (2z - 1)^2 - 4\mu\nu_T z\sqrt{\nu_T^2 z - 1}\sqrt{z - 1} \stackrel{\text{def}}{=} \theta(z)$$

because $\mu\nu_T^2 = 1$. We have, for $z \geq 1$,

$$\theta'(z) = 4\left[(2z - 1)\right.$$
$$\left. -\mu\nu_T\left(\sqrt{z - 1}\sqrt{\nu_T^2 z - 1} + z\frac{\nu_T^2\sqrt{z - 1}}{2\sqrt{\nu_T^2 z - 1}} + z\frac{\sqrt{\nu_T^2 z - 1}}{2\sqrt{z - 1}}\right)\right] < 0 \quad (3.72)$$

(Remark that $z \geq 1$ ensures $z - 1 \leq \mu\nu_T\sqrt{z - 1}\sqrt{\nu_T^2 z - 1}$ and $z < \mu\nu_T\frac{z}{2}\left(\frac{\nu_T^2\sqrt{z-1}}{\sqrt{\nu_T^2 z-1}} + \frac{\sqrt{\nu_T^2 z-1}}{\sqrt{z-1}}\right)$).

Since $\theta(1) = 1$ and $\theta(z) \simeq -2z(1 - \mu)$, when $z \to +\infty$, for $\xi \in [\nu_T, +\infty[$, $R(\xi)$ decreases strictly from 1 to $-\infty$. Therefore, $R(\xi)$ does have an unique root $\xi = \nu_R \in]\nu_T, +\infty[$. The Rayleigh velocity is defined by

$$\nu_R = \frac{c_L}{c_R}. \quad (3.73)$$

Since $S(\xi) = R(\xi) + \rho\frac{\sqrt{\xi^2 - 1}}{\sqrt{\xi^2 - \nu_0^2}}$ and since the function $\xi \in]\nu_0, +\infty[\mapsto \rho\frac{\sqrt{\xi^2 - 1}}{\sqrt{\xi^2 - \nu_0^2}}$ decreases from $+\infty$ to ρ, the function $\xi \in]\nu_0, +\infty[\mapsto S(\xi)$ is strictly decreasing from $+\infty$ to $-\infty$. Therefore, $S(\xi) = 0$ does have an unique root $\xi = \nu_S \in]\nu_0, +\infty[$. The *Scholte-Stoneley* velocity is defined by

$$\nu_S = \frac{c_L}{c_S}. \quad (3.74)$$

■

Let us suppose $c_R > c_0$, i.e. $\nu_R < \nu_0$. For small values of $\rho > 0$, we have one complex root $\nu_R(\rho)$ close to ν_R, of the equation $S(\xi) = 0$. For ξ in a neighborhood of ν_R, we have $S(\xi) = R(\xi) - i\rho\frac{\sqrt{\xi^2-1}}{\sqrt{\nu_0^2-\xi^2}}$ and by an expansion to the first order, we obtain $(R'(\nu_R) < 0)$

$$\nu_R(\rho) = \nu_R - i\frac{\rho}{|R'(\nu_R)|}\frac{\sqrt{\nu_R^2-1}}{\sqrt{\nu_0^2-\nu_R^2}} + O(\rho^2). \qquad (3.75)$$

The singularities of the matrix $dm(\xi)$ are pictured on Fig. 3.2

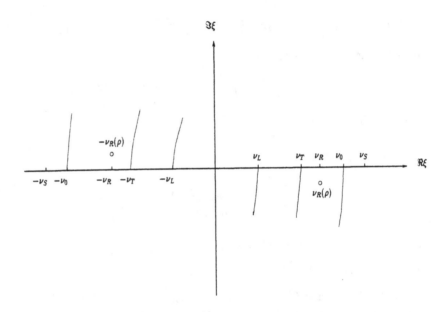

Fig. 3.2. The singularities of the matrix $dm(\xi)$

The points $\xi = (\pm 1 = \pm\nu_L), \pm\nu_T, \pm\nu_0$ are branch points. On Fig.3.2, we have chosen the branch lines from these points in order to make apparent the position of the Rayleigh roots $\pm\nu_R(\rho)$ on the Riemann surface of $dm(\xi)$. The holomorphic extension of $dm(\xi)$ from the domain V is non invertible at $\pm\nu_S$, $\pm\nu_R(\rho)$. We do not compute further the other singularities of $dm(\xi)$ on its Riemann surface, because it is not useful in the sequel.

Lemma 3.5 (Properties of the matrix $dm(z)$).

(i) The matrix $dm(z)$ is non singular for $z \in V$.

(ii) The matrix $dm(z)$ is bounded on each axis $z = \xi e^{i\theta}, \theta \in]0, \pi[.$

(iii)The matrix $dm^{-1}(z)$ is bounded on V outside a neighborhood of $\pm\nu_S$.

Proof.

(i) For $\theta \in]0, \pi[$, the non-existence of an interface-wave solution to the system (S_θ) (cf Lemma 3.3), is equivalent to the non-existence of a non-zero vector $[\hat{\alpha}_\xi, \hat{\beta}_\xi, \hat{\gamma}_\xi]^T$ solution of the equation (3.61). Hence, the matrix $dm(e^{i\theta}\xi)$ is non singular for each $\xi \in \mathbb{R}$. In addition, we just see in Lemma 3.4 (ii), that $\det(dm(\xi)) \neq 0$ for $\xi \in [-1, 1]$. This gives the result.

(ii) The result follows from the definition of the coefficients A, B, C, D and from the asymptotic expression of $Q(\xi e^{i\theta}) = e^{2i\theta}[\varsigma_L^\theta(\xi)\varsigma_T^\theta(\xi) + \xi^2]$ for large values of $|\xi|$

$$Q(\xi e^{i\theta}) = e^{2i\theta}\left(\xi^2 - \sqrt{\xi^2 - \nu_L^2 e^{-2i\theta}}\sqrt{\xi^2 - \nu_T^2 e^{-2i\theta}}\right) \simeq \frac{1}{2}(\nu_L^2 + \nu_T^2) = \frac{1}{2}(1 + \nu_T^2) \tag{3.76}$$

which proves that $Q(\xi e^{i\theta})$ remains bounded.

(iii) For large values of $|\xi|$, $\Delta(\xi e^{i\theta}) \simeq -(1 + \mu)(3 + \mu)$, therefore we have only to check that $dm^{-1}(z)$ remains bounded for z close from $\pm\nu_{L,T,0}$. This results follows from (3.64) where a simplification occurs between $\frac{1}{\Delta}$ and the singularity of the coefficients. ∎

3.3. Some Properties of the Operators DM, TM

Let us introduce two functional spaces useful in the sequel.

• We call H^+ the space of the functions $f(\xi)$, analytic in the lower half-plane and uniformly L^2 on the horizontal lines, *i.e.*

$$\sup_{c>0} \int_\mathbb{R} |f(\xi - ic)|^2 d\xi < +\infty. \tag{3.77}$$

The theorem of Paley-Wiener just states that these functions $f(\xi)$ are of the form

$$f(\xi) = \int_0^{+\infty} e^{-ix\xi} g(x)dx, \quad g \in L^2(\mathbb{R}^+, \mathbb{C}). \tag{3.78}$$

The space H^+ is a Hilbert space for the L^2-norm of $L^2(\mathbb{R}^+, \mathbb{C})$.

• We call $\mathcal{H}$ the space of the function analytic in $U = \mathbb{C}\backslash] - \infty, -\nu_L = 1]$ and such that $f(\xi e^{i\alpha}) \in H^+$ for each $\alpha \in]0, \pi[$, (Fig 3.3).

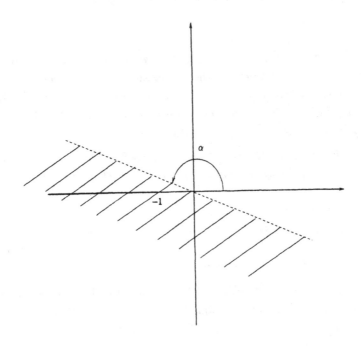

Fig. 3.3. $f \in \mathcal{H} \Longleftrightarrow f(\xi e^{i\alpha}) \in H^+, \ \alpha \in]0, \pi[$

Lemma 3.6. *For $\theta \in]0, \pi[$, the operator*

$$f(\xi) \mapsto DM^\theta(f)(\xi) = \int_{\Gamma_0} DM^\theta(\xi, \zeta) \cdot f(\zeta) d\zeta$$

is bounded on $(H^+)^3$.

Proof. We have $DM^\theta(\xi, \zeta) = \frac{1}{2i\pi} \frac{1}{\xi - \zeta} dm(e^{i\theta}\zeta)$. Hence, Lemma 3.6 results from Lemma 3.5 (ii) and from the boundedness of the Hilbert projector from $L^2(\mathbb{R})$ onto H^+, (cf Appendix). ∎

Lemma 3.7. *For $\theta \in]0, \pi[$, the operator*

$$f(\xi) \mapsto TM^\theta(f)(\xi) = \int_{\Gamma_0} TM^\theta(\xi, \zeta) \cdot f(\zeta) d\zeta$$

is bounded on $(H^+)^3$. In addition, there exist three constants $c_0(\theta), c_1(\theta), c_2(\theta)$ such that $TM^\theta(f)(\xi)$ can be extended analytically to the domain $W = \{ \text{Im} \, \xi < c_0(\theta) + c_1(\theta) | \text{Re} \, \xi | \}$ and satisfies $TM^\theta(f)(\xi e^{i\alpha}) \in (H^+)^3$ for $|\alpha| \leq c_2(\theta)$.

Proof. Due to the Hardy inequality, (cf Appendix), it is sufficient to check that $TM^\theta(\xi, \zeta)$ is analytic in $\xi \in W$ for $\zeta \in \mathbb{R}$ and that there exists a constant K such that

$$|TM^\theta(\xi, \zeta)| \leq K(1 + |\xi| + |\zeta|)^{-1}. \tag{3.79}$$

The kernel $TM^\theta(\xi, \zeta)$ can be set in the form

$$TM^\theta(\xi, \zeta) = \frac{1}{2i\pi}\Big[TM^{\theta,0}(\xi, \zeta) + \sin\varphi TM^{\theta,1}(\xi, \zeta) + \cos\varphi TM^{\theta,2}(\xi, \zeta)$$
$$+ \sin 2\varphi TM^{\theta,3}(\xi, \zeta) + \cos 2\varphi TM^{\theta,4}(\xi, \zeta)\Big] \tag{3.80}$$

where the matrices $TM^{\theta,j}(\xi, \zeta)$, $j = 0, 1, 2, 3, 4$ are given by

$$TM^{\theta,0} = \begin{pmatrix} 0 & 0 & 0 \\ -D_L^\theta \frac{\zeta}{\zeta_L^\theta}(1-\mu) & -D_L^\theta(1-\mu) & D_0^\theta \frac{\rho}{\zeta_0} \\ 0 & 0 & 0 \end{pmatrix} \tag{3.81}$$

$$TM^{\theta,1} = \begin{pmatrix} 0 & 0 & 0 \\ 0 & 0 & 0 \\ -D_L^\theta \frac{\zeta^2}{\zeta_L^\theta} - D_T^\theta \zeta_T^\theta & (D_T^\theta - D_L^\theta)\zeta & D_0^\theta \frac{\zeta}{\zeta_0} \end{pmatrix} \tag{3.82}$$

$$TM^{\theta,2} = \begin{pmatrix} 0 & 0 & 0 \\ 0 & 0 & 0 \\ (D_L^\theta - D_T^\theta)\zeta & D_L^\theta \zeta_L^\theta + D_T^\theta \frac{\zeta^2}{\zeta_T^\theta} & -D_0^\theta \end{pmatrix} \tag{3.83}$$

$$TM^{\theta,3} = \mu \begin{pmatrix} D_L^\theta\left(\zeta\zeta_L^\theta - \frac{\zeta^3}{\zeta_L^\theta}\right) - 2D_T^\theta\zeta\zeta_T^\theta & D_L^\theta(\zeta_L^2 - \zeta^2) + 2D_T^\theta\zeta^2 & 0 \\ 2D_L^\theta\zeta^2 - D_T^\theta(\zeta^2 - \zeta_T^2) & 2D_L^\theta\zeta\zeta_L^\theta + D_T^\theta\left(\frac{\zeta^3}{\zeta_T^\theta} - \zeta\zeta_T^\theta\right) & 0 \\ 0 & 0 & 0 \end{pmatrix} \tag{3.84}$$

$$TM^{\theta,4} = \mu \begin{pmatrix} 2D_L^\theta\zeta^2 + D_T^\theta((\zeta_T^\theta)^2 - \zeta^2) & 2D_L^\theta\zeta\zeta_L^\theta + D_T^\theta\left(\frac{\zeta^3}{\zeta_T^\theta} - \zeta\zeta_T^\theta\right) & 0 \\ D_L^\theta\left(\frac{\zeta^3}{\zeta_L^\theta} - \zeta\zeta_L^\theta\right) + 2D_T^\theta\zeta\zeta_T^\theta & D_L^\theta(\zeta^2 - (\zeta_L^\theta)^2) - 2D_T^\theta\zeta^2 & 0 \\ 0 & 0 & 0 \end{pmatrix} \tag{3.85}$$

We have $\mathrm{Im}(\cos\varphi\zeta + \sin\varphi\zeta^{\theta}_{L,T,0}) = \sin\varphi\,\mathrm{Im}(\nu^2_{L,T,0}e^{-2i\theta} - \zeta^2)^{1/2} \geq K(1 + |\zeta|)$ for $\zeta \in \mathbb{R}$ (K constant), due to the choice of the determination of $(\nu^2_{L,T,0}e^{-2i\theta} - \zeta^2)^{1/2}$. Therefore, for sufficiently small coefficients $c_j(\theta)$, we have for $\xi \in \mathbb{R}$, $\zeta \in V$

$$|D^{\theta}_*| \leq K(1 + |\xi| + |\zeta|)^{-1} \tag{3.86}$$

As an example, we have for the coefficient $TM^{\theta,3}(1,1)$

$$TM^{\theta,3}(1,1) = e^{2i\theta}\Big\{ D^{\theta}_L\Big(\zeta\zeta^{\theta}_L - \frac{\zeta^3}{\zeta^{\theta}_L}\Big) - 2D^{\theta}_T\zeta\zeta^{\theta}_T \Big\} \tag{3.87}$$

$$= e^{2i\theta}\Big\{ D^{\theta}_L D^{\theta}_T \frac{\zeta}{\zeta^{\theta}_L}\Big[\frac{e^{-2i\theta}}{D^{\theta}_T} - 2\Big(\frac{\zeta^2}{D^{\theta}_T} + \frac{\zeta^{\theta}_T\zeta^{\theta}_L}{D^{\theta}_L}\Big)\Big]\Big\}.$$

Moreover

$$\frac{\zeta^2}{D^{\theta}_T} + \frac{\zeta^{\theta}_T\zeta^{\theta}_L}{D^{\theta}_L} = (\xi - \cos\varphi\zeta)Q(\zeta e^{i\theta}) - \sin\varphi\zeta^{\theta}_T e^{-2i\theta}. \tag{3.88}$$

The result follows from the boundedness of $Q(\zeta e^{i\theta})$ and from (3.86). The other terms are handled similarly. ∎

Remark: Lemma 3.7 could be proved without computations, assuming general results on the elliptic operators. The cancellation properties that we just check, give an a posteriori verification of the computations leading to the formulas (3.81-3.85).

The rest of this paragraph is devoted to the structure of the two functions

$$\xi \mapsto \int_{\Gamma_0} DM(\xi,\zeta)\cdot\frac{v}{\zeta - z}d\zeta \ ; \ \xi \mapsto \int_{\Gamma_0} TM(\xi,\zeta)\cdot\frac{v}{\zeta - z}d\zeta \tag{3.89}$$

where $v \in \mathbb{C}^3$ and $z \in U = \mathbb{C}\backslash] - \infty, -\nu_L = 1]$, $\mathrm{Im}\,z \geq 0$, $z \notin \{\nu_L, \nu_T, \nu_0\}$. Let us first recall the definition of the translation operators $T_*(z)$ introduced in Sect.2.4. The domain Ω_* is defined by

$$\Omega_* = \{\xi = \nu_* \cos\theta, \theta \in \mathcal{D}, \mathrm{Re}\,\theta < \pi - \theta\} \tag{3.90}$$

where $\mathcal{D} = \{\theta \in \mathbb{C}, 0 < \mathrm{Re}\,\theta < \pi\} \cup \{\theta = -it, t \geq 0\} \cup \{\theta = \pi + it, t \geq 0\}$. The operator T_* is defined for $\xi = \nu_* \cos\theta \in \Omega_*$ by $T_*(\xi) = \nu_* \cos(\theta + \varphi)$. The geometric construction of $T_\nu(\xi)$ is obtained as follows. The point $\xi = \nu\cos(\theta_1 + i\theta_2)$ is located at the intersection of the ellipse $\mathcal{E}_{\nu,\xi}$ centered at the origin and of semi-axis $\nu\cosh\theta_2$, $\nu|\sinh\theta_2|$ and of the hyperbola $\{z = \nu\cos\theta, \mathrm{Re}\,\theta = \theta_1\}$. The point $T_\nu(\xi) = \nu\cos(\theta+\varphi)$ is located at the intersection of the same ellipse and of the hyperbola $\{z = \nu\cos\theta, \mathrm{Re}\,\theta = \theta_1 + \varphi\}$. In addition, if $\xi = \nu\cos\theta$ is on the real axis, we have two cases

- $\nu > \xi$. In this case, $\theta \in]0, \pi - \varphi[$, hence the point $T_\nu(\xi) = \nu\cos(\theta + \varphi)$ is still real (Fig. 3.4).

- $\nu \leq \xi$. In this case $\theta_2 < 0$, hence $T_\nu(\xi) \in \{\xi / \operatorname{Im}\xi > 0\}$ (Fig. 3.5).

Lemma 3.8. *Suppose that* $v \in \mathbb{C}^3$, $z \in \mathbb{U}$, $z \neq \{\nu_L, \nu_T, \nu_0\}$, $\operatorname{Im} z \geq 0$. *Let us define the functions* $F_D(z, v)$ *and* $F_T(z, v)$ *on the domain* $\operatorname{Im}\xi < 0$ *by the following integrals (the contour Γ_0 is below z for $z \in \mathbb{R}$)*

$$F_D(z,v)(\xi) = \int_{\Gamma_0} DM(\xi,\zeta) \cdot \frac{v}{\zeta - z} d\zeta \; ; \; F_T(z,v)(\xi) = \int_{\Gamma_0} TM(\xi,\zeta) \cdot \frac{v}{\zeta - z} d\zeta$$

can be decomposed in the form

$$F_D(z,v)(\xi) = \frac{1}{\xi - z} dm(z) \cdot v + D_p(z,v)(\xi) \tag{3.91}$$

$$F_T(z,v)(\xi) = \sum_{*\in\{L,T,0\}} \frac{tm_*(z) \cdot v}{\xi - T_*(z)} \, 1(z \in \Omega_*) + T_p(z,v)(\xi) \tag{3.92}$$

where the functions $D_p(z,v)(\xi)$, $T_p(z,v)(\xi)$ *are in* $\mathcal{H}^3$.

Proof. The function F_T is defined on the half-plane $\operatorname{Im}\xi < 0$ because of Lemma 3.7 and of $(xe^{i\alpha} - z)^{-1} \in H^+$ for $0 < \alpha < \alpha_0$. The decompositons (3.91), (3.92) result from the Cauchy formula and from the residue formula. For $F_D(z,v)$ we deform the contour Γ_0 onto the contour Γ_1 pictured on Fig. 3.6. This yields

$$D_p(z,v)(\xi) = \int_{\Gamma_1} DM(\xi,\zeta) \cdot \frac{v}{\zeta - z} d\zeta. \tag{3.93}$$

Moreover, Lemma 3.5 (ii) ensures the estimates

$$|DM(\xi e^{i\alpha}, \zeta)| = \left| \frac{1}{2i\pi} \frac{1}{\xi e^{i\alpha} - \zeta} dm(\zeta) \right| \leq \frac{C}{1 + |\xi| + |\zeta|} \tag{3.94}$$

for $\xi \in \mathbb{R}$, $\zeta \in \Gamma_1$. The fact that $D_p(z,v)(\xi) \in \mathcal{H}$ results from the Hardy inequality, (cf Appendix). We consider now $T_p(z,v)(\xi)$. The kernel $TM(\xi,\zeta)$ can be written (see (3.34))

$$TM(\xi,\zeta) = \frac{1}{2i\pi}\Big(D_L(\xi,\zeta)tm_L(\zeta) + DM_T(\xi,\zeta)tm_T(\zeta) + DM_0(\xi,\zeta)tm_0(\zeta)\Big) \tag{3.95}$$

where $tm_*(\zeta)$ is given by (3.24-3.26).

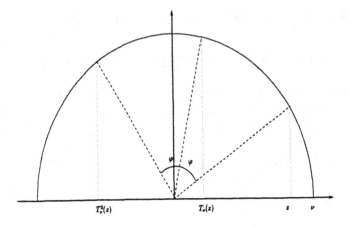

Fig. 3.4. The iterates $T_\nu^n(z)$ of z, when $\nu > \xi$

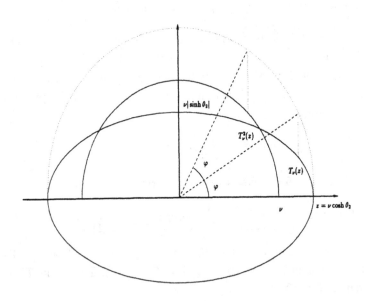

Fig. 3.5. The iterates $T_\nu^n(z)$ of z, when $\nu \leq \xi$

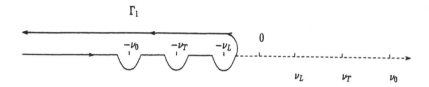

Fig. 3.6. The contour Γ_1

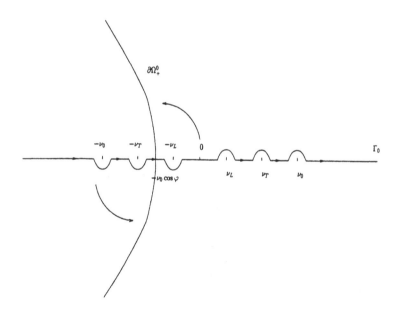

Fig. 3.7 Deformation of the contour Γ_0 onto the contour $\partial\Omega_0^+$

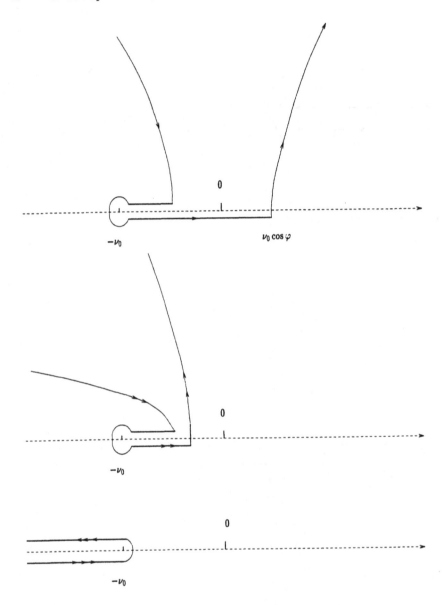

Fig. 3.8. Deformation of the contour $\zeta \mapsto \zeta \cos \varphi + \sin \varphi \zeta_0$ corresponding to the deformation of the contour $\zeta \in \Gamma_0$ onto $\partial \Omega_0^+$

We decompose $F_T(z, v)(\xi)$ into two parts

$$F_T(z, v)(\xi) = G(\xi) + H(\xi) \tag{3.96}$$

$$G(\xi) = \int_{\Gamma_0} \frac{1}{2i\pi} \Big(D_L(\xi, \zeta) tm_L(\zeta) + D_T(\xi, \zeta) tm_T(\zeta) \Big) \cdot \frac{v}{\zeta - z} d\zeta \tag{3.97}$$

$$H(\xi) = \int_{\Gamma_0} \frac{1}{2i\pi} \Big(D_0(\xi, \zeta) tm_0(\zeta) \cdot \frac{v}{\zeta - z} d\zeta. \tag{3.98}$$

For $H(\xi)$, the contour Γ_0 id deformed onto the contour $\partial\Omega_+^0$ (Fig. 3.7). The deformation of $\cos\varphi\zeta + \sin\varphi\zeta_0(\zeta)$ is pictured on Fig. 3.7. If $z \in \Omega_0^+$, the residue is

$$D_0(\xi, z) tm_0(z) \cdot v = \frac{1}{\zeta - T_0(z)} tm_0(z) \cdot v. \tag{3.99}$$

(If $z \in \partial\Omega_+^0$, the final contour will be at the right of z).

The part $G(\xi)$ is more subtle because $tm_L(\zeta)$, $tm_T(\zeta)$ are quadratic in ζ, for $|\zeta|$ large. Thus, we cannot break the term $D_L tm_L + D_T tm_T$. However the two contour $\partial\Omega_L^+$, $\partial\Omega_T^+$ are homothetic and form asymptotic hyperbolic branchs for $|\zeta|$ large. This allows to write again

$$G(\xi) = \Big\{ \frac{1}{2i\pi} \int_{\partial\Omega_L^+} D_L(\xi, \zeta) tm_L(\zeta) \cdot \frac{v}{\zeta - z} d\zeta$$
$$+ \frac{1}{2i\pi} \int_{\partial\Omega_T^+} D_T(\xi, \zeta) tm_T(\zeta) \cdot \frac{v}{\zeta - z} d\zeta \Big\} + \frac{w_L}{\xi - T_L(z)} + \frac{w_T}{\xi - T_T(z)}. \tag{3.100}$$

We check that the function inside the braces is in $\mathcal{H}^3$. The integrals are decomposed into two parts according to

$$(i) \quad |\zeta| \le M, \qquad (ii) \quad |\zeta| > M \tag{3.101}$$

where the contour $\partial\Omega_L^+$, $\partial\Omega_T^+$ are matched for $|\zeta|$ large. For (i), we use the estimate, true for $\xi \in \mathbb{R}$, $\zeta \in \partial\Omega_{L,T}^+$

$$\Big| D_{L,T}(\xi e^{i\alpha}, \zeta) tm_{L,T}(\zeta) \cdot \frac{v}{\zeta - z} \Big| \le \frac{CM^2}{1 + |\xi|} \tag{3.102}$$

and for (ii)

$$\Big| D_L(\xi e^{i\alpha}, \zeta) tm_L(\zeta) + D_T(\xi e^{i\alpha}, \zeta) tm_T(\zeta) \Big| \le \frac{C}{1 + |\xi| + |\zeta|}. \tag{3.103}$$

Finally, we get the conclusion of the lemma with $T_p(z, v)(\xi)$ defined by

$$T_p(z,v)(\xi) = \frac{1}{2i\pi} \int_{\partial\Omega_0^+} D_0(\xi,\zeta) tm_0(\zeta) \cdot \frac{v}{\zeta - z} d\zeta$$

$$+ \frac{1}{2i\pi} \left\{ \int_{\partial\Omega_L^+} D_L(\xi,\zeta) tm_L(\zeta) \cdot \frac{v}{\zeta - z} d\zeta \right. \tag{3.104}$$

$$\left. + \int_{\partial\Omega_T^+} D_T(\xi,\zeta) tm_T(\zeta) \cdot \frac{v}{\zeta - z} d\zeta \right\}$$

∎

Remark: We can compute explicitly the function $F_T(z,v)(\xi)$ with the help of usual functions. However the final form is complicated and does not bring any additional information.

3.4. Decomposition of the Spectral Function

The aim of this paragraph is to explain how the decomposition of the spectral function in the form $\Sigma(\xi) = y(\xi) + X(\xi)$ is obtained. The first one, $y(\xi)$, is a meromorphic function having its poles in the set $\{z \in \mathbb{C}, \operatorname{Im} z \geq 0, z \notin \{\nu_0, \nu_L, \nu_T\}\}$. The second one, $X(\xi)$, will be in $\mathcal{H}^3$.

We begin with the explanation of this decomposition. Recall that the system to solve with unknown $\Sigma(\xi) = (\Sigma_1(\xi), \Sigma_2(\xi))$ in $\hat{A}^6$ is (cf § 3.1, (3.37)

$$\begin{cases} DM \cdot \Sigma_1(\xi) + TM \cdot \Sigma_2(\xi) = \dfrac{W_1}{\xi - Z_1} \\[3mm] TM \cdot \Sigma_1(\xi) + DM \cdot \Sigma_2(\xi) = \dfrac{W_2}{\xi - Z_2} \end{cases}, \quad \operatorname{Im} \xi < 0 \tag{3.105}$$

where $\theta_{in} \in]-(\pi - \varphi), \pi[$ is the incidence angle of the incoming wave, $Z_1 = \nu_0 \cos\theta_{in}$, $Z_2 = \nu_0 \cos(\theta_{in} + \varphi)$. and $W_1, W_2 \in \mathbb{C}^3$ are given by (3.32). Suppose that the incident wave illuminates only the first face of the wedge, $(\theta_{in} \in]0, \pi - \varphi[)$. We define at the first step $X_j(\xi)$, $j = 1, 2$ by

$$\Sigma_j(\xi) = \frac{V_j}{\xi - Z_j} + X_j(\xi), \qquad j = 1, 2 \tag{3.106}$$

where $V_j \in \mathbb{C}^3$ is unknown. By using Lemma 3.8, $DM \cdot \Sigma_j$ can be written

$$DM \cdot \Sigma_j(\xi) = \frac{dm(Z_j) \cdot V_j}{\xi - Z_j} + D_p(V_j, Z_j)(\xi) + DM \cdot X_j(\xi). \tag{3.107}$$

Choosing $V_j = dm^{-1}(Z_j)W_j$, we deduce from (3.105) the following system with unknown (X_1, X_2)

$$
\begin{cases}
DM \cdot X_1 + TM \cdot X_2 = -\left(TM \cdot \dfrac{V_2}{\zeta - Z_2}\right)(\xi) + D_p(V_1, Z_1)(\xi) \\[4mm]
TM \cdot X_1 + DM \cdot X_2 = -\left(TM \cdot \dfrac{V_1}{\zeta - Z_1}\right)(\xi) + D_p(V_2, Z_2)(\xi).
\end{cases}
\tag{3.108}
$$

We deduce from Lemma 3.8, that

$$
\left(TM \cdot \frac{V_j}{\zeta - Z_j}\right)(\xi) = \sum_{*\in\{L,T,0\}} \frac{tm_*(Z_j) \cdot V_j}{\xi - T_*(Z_j)} \; \mathbf{1}(Z_j \in \Omega_*^+) + T_p(V_j, Z_j)(\xi)
\tag{3.109}
$$

Again, we extract from $X_j(\xi)$ the singular function

$$
\sum_{*\in\{L,T,0\}} \frac{tm_*(Z_j) \cdot V_j}{\xi - T_*(Z_j)} \; \mathbf{1}(Z_j \in \Omega_*^+)
$$

and we redefine $X_1(\xi), X_2(\xi)$ at the second step by

$$
\begin{cases}
\Sigma_1(\xi) = \dfrac{V_1}{\xi - Z_1} \\[3mm]
\qquad + \displaystyle\sum_{*\in\{L,T,0\}} \dfrac{dm^{-1}(T_*(Z_2)) \cdot tm_*(Z_2) \cdot V_2}{\xi - T_*(Z_2)} \; \mathbf{1}(Z_2 \in \Omega_*^+) + X_1(\xi) \\[5mm]
\Sigma_2(\xi) = \dfrac{V_2}{\xi - Z_2} \\[3mm]
\qquad + \displaystyle\sum_{*\in\{L,T,0\}} \dfrac{dm^{-1}(T_*(Z_1)) \cdot tm_*(Z_1) \cdot V_1}{\xi - T_*(Z_1)} \; \mathbf{1}(Z_1 \in \Omega_*^+) + X_2(\xi)
\end{cases}
\tag{3.110}
$$

It is now clear how to proceed in order to define the decomposition $\Sigma = y + X$. Let us introduce the definition of the functions $y_1(\xi)$, $y_2(\xi)$. We call $P^j = Z^j(\xi_1, \xi_2)$ the finite subsets of $\mathbb{C}$ defined in (2.34), and

$$
P^j(\mathbb{U}) = \{z \in P^j, z \in \mathbb{U} = \mathbb{C}\setminus] - \infty, -\nu_L]\}.
\tag{3.111}
$$

Under the hypothesis (H) (cf § 2.5) on the incidence angle θ_{in}, we have

$$
P^j(\mathbb{U}) \cap \{\nu_L, \nu_T, \nu_0\} = \emptyset.
\tag{3.112}
$$

We seek the functions $y_1(\xi)$, $y_2(\xi)$, meromorphic in $\mathbb{U}$, in the form

$$
y_j(\xi) = \sum_k \frac{a_j^k}{\xi - z_k^j}, \quad a_j^k \in \mathbb{C}^3, \quad z_k^j \in P^j(\mathbb{U}).
\tag{3.113}
$$

Lemma 3.9. *There exist two functions $y_1(\xi)$, $y_2(\xi)$ of the form (3.113) such that*

$$
\begin{cases}
\dfrac{W_1}{\xi - Z_1} - DM \cdot y_1 - TM \cdot y_2 = u_1(\xi) \in \mathcal{H}^3 \\[2mm]
\dfrac{W_2}{\xi - Z_2} - DM \cdot y_2 - TM \cdot y_1 = u_2(\xi) \in \mathcal{H}^3.
\end{cases}
\tag{3.114}
$$

Proof. For $* \in \{L, T, 0\}$ and $z \in \Omega_* \setminus \{\nu_L, \nu_T, \nu_0, \nu_S\}$ such that $T_*(z) \neq \nu_*$, we define the 3×3 matrices $\tilde{t}_*(z)$ by

$$
\tilde{t}_*(z) = dm^{-1}(T_*(z))tm_*(z).
\tag{3.115}
$$

We deduce easily from the formulas (3.10-3.12) that (recall that we note $S(\xi) = \delta(\xi)$, where $S(\xi)$ is defined by (3.65)).

- for $* = 0$, $z = \nu_0 \cos\chi$, $T_0(z) = \nu_0 \cos\psi$, with $\psi = \theta + \varphi$

$$
\tilde{t}_0(z) = \frac{\sin\psi}{\sin\chi}[0, 0; z_0], \quad z_0 = \begin{bmatrix} 0 \\ 0 \\ 1 \end{bmatrix}
\tag{3.116}
$$

- for $* = L$, $z = \nu_L \cos\chi$, $T_L(z) = \nu_L \cos\psi$, with $\psi = \theta + \varphi$

$$
\tilde{t}_L(z) = \frac{\sin\psi}{\sin\chi}\frac{1}{\delta(\xi)}\left[\cos\chi z_L(\xi), -\sin\chi z_L(\xi), 0\right]
\tag{3.117}
$$

$$
\delta(\xi) = 1 - 4\mu\xi^2 + 4\mu^2\xi^2 Q(\xi) + \rho\frac{\zeta_L(\xi)}{\zeta_0(\xi)}
$$

$$
z_L(\xi) = \begin{bmatrix} -\frac{\xi}{Q(\xi)}(\delta(\xi) + (1 - 2\mu Q(\xi))(4\mu\xi^2 - 2)) \\ -\frac{\zeta_T(\xi)}{Q(\xi)}(\delta(\xi) + 4\mu\xi^2 - 2) \\ 4\mu(\xi^2 - 2) \end{bmatrix}
$$

- for $* = T$, $z = \nu_T \cos\chi$, $T_T(z) = \nu_T \cos\psi$, with $\psi = \theta + \varphi$

$$
\tilde{t}_T(z) = \frac{\sin\psi}{\sin\varphi}\frac{1}{\delta(\xi)}\left[\sin\theta z_T, \cos\chi z_T, 0\right]
\tag{3.118}
$$

$$
z_T(\xi) = \begin{bmatrix} \nu_T\frac{\zeta_L(\xi)}{Q(\xi)}(\delta(\xi) + 4\mu\xi^2(1 - 2\mu Q(\xi))) \\ -\nu_T\frac{\xi}{Q(\xi)}(\delta(\xi) - 4\mu\zeta_L(\xi)\zeta_T(\xi)) \\ -4\mu\nu_T\xi\zeta_L(\xi) \end{bmatrix}.
$$

In particular, we note that

$$
\tilde{t}_*(T_*(z)) \circ \tilde{t}_0(z) = 0, \quad * = L, T
\tag{3.119}
$$

because the third column of the matrices $\tilde{t}_{L,T}$ are zero. The sets $\mathcal{Z}^j(Z_1, Z_2) = \bigcup_{\ell \geq 0} \mathcal{Z}_\ell^j(Z_1, Z_2)$ are built by recurrence on the generation index ℓ by the formulas (2.34-2.35) and we define $\mathcal{P}^j(\mathbb{U}) = \bigcup_{\ell \geq 0} \mathcal{P}_\ell^j(\mathbb{U})$ with $\mathcal{P}_\ell^j(\mathbb{U}) = \mathcal{Z}_\ell^j(Z_1, Z_2) \cap \mathbb{U}$. We seek $y_1(\xi)$, $y_2(\xi)$ in the form

$$y_j(\xi) = \sum_{\ell \geq 0} y_{j,\ell}(\xi), \quad y_{j,\ell}(\xi) = \sum_k \frac{a_{j,\ell}^k}{\xi - z_{j,\ell}^k}, \quad z_{j,\ell}^k \in \mathcal{P}_\ell^j(\mathbb{U}). \tag{3.120}$$

Lemma 3.9 follows now from Lemma 3.8 by defining the values $a_{j,\ell}^k$ by recurrence on ℓ as

• $\ell = 0$: For $j = 1, 2$, $a_{j,0} = dm(Z_j)^{-1} \cdot W_j$

• $\ell \Rightarrow \ell + 1$: If $z_{j,\ell+1}^k = z \in \mathcal{P}_{\ell+1}^j(\mathbb{U})$ is equal to $T_*(z')$ with $z' = z_{j',\ell}^{k'}$, ($j' = 2$ if $j = 1$ and $j' = 1$ if $j = 2$) in (2.34-2.35) then $a_{j,\ell+1}^k$ is defined by

$$a_{j,\ell+1}^k = -\tilde{t}_*(z') \cdot a_{j',\ell}^{k'}. \tag{3.121}$$

∎

We call now $X_1(\xi), X_2(\xi)$ the functions, defined on $\mathrm{Im}\,\xi < 0$

$$X_j(\xi) = \Sigma_j(\xi) - y_j(\xi). \tag{3.122}$$

Subtracting the vector $\begin{bmatrix} DM \cdot y_1 + TM \cdot y_2 \\ TM \cdot y_1 + DM \cdot y_2 \end{bmatrix}$ from each side of (3.105), we obtain that (3.105) is equivalent to the system (3.123) with unknown $(X_1, X_2) \in \hat{\mathcal{A}}^6$ and right-hand side $(u_1, u_2) \in \mathcal{H}^6$ given by

$$\begin{cases} DM \cdot X_1(\xi) + TM \cdot X_2(\xi) = u_1(\xi) \\ TM \cdot X_1(\xi) + DM \cdot X_2(\xi) = u_2(\xi). \end{cases} \tag{3.123}$$

Theorems 1 and 2, proved in § 4, state that (3.123) has an unique solution in $\mathcal{H}^6 \subset \hat{\mathcal{A}}^6$ and that this solution is even the sole solution in $\hat{\mathcal{A}}^6$.

3.5. The Functional Equation for the Spectral Function

The aim of this paragraph is to establish a functional identity for the spectral function. This identity is useful both theoretically and numerically for the evaluation of the boundary values of the spectral function on the real axis from the lower half plane

$$\lim_{\varepsilon \to 0^+} \Sigma(\xi - i\varepsilon), \quad \xi \in \mathbb{R}.$$

Suppose we know that $(\Sigma_1, \Sigma_2) \in \hat{\mathcal{A}}^6$ is solution of the system (3.105) (see Sect.4 for the proof of existence and uniqueness and Def.2.1 for the definition of $\hat{\mathcal{A}}$).

$$\begin{cases} \int_{\Gamma_0} DM(\xi, \varsigma) \cdot \Sigma_1(\varsigma)d\varsigma + \int_{\Gamma_0} TM(\xi, \varsigma) \cdot \Sigma_2(\varsigma)d\varsigma = \dfrac{W_1}{\xi - Z_1} \\ \int_{\Gamma_0} TM(\xi, \varsigma) \cdot \Sigma_1(\varsigma)d\varsigma + \int_{\Gamma_0} DM(\xi, \varsigma) \cdot \Sigma_2(\varsigma)d\varsigma = \dfrac{W_2}{\xi - Z_2}. \end{cases} \quad (3.124)$$

We deform in (3.124) the contour Γ_0 onto the contour Γ_2 pictured on Fig 3.9.

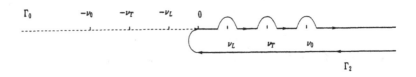

Fig. 3.9. The contours Γ_0 and Γ_2

By the Cauchy formula, we have for the terms in DM, for $\xi \in \mathbb{C}$, $\text{Im}\,\xi < 0$

$$\int_{\Gamma_0} DM(\xi, \varsigma) \cdot f(\varsigma)d\varsigma = \int_{\Gamma_2} DM(\xi, \varsigma) \cdot f(\varsigma)d\varsigma + dm(\xi) \cdot f(\xi). \quad (3.125)$$

For the terms in TM, we introduce the domain

$$\Omega_*^- = \left\{ \xi \in \mathbb{C}, \text{Im}\,\xi < 0, \xi = \nu_* \cos\theta, \varphi < \text{Re}\,\theta < \pi \right\}. \quad (3.126)$$

For $\xi = \nu_* \cos\theta \in \Omega_*^-$, we define the operator inverse of the translation T_* by

$$T_*^{-1}(\nu_* \cos\theta) = \nu_* \cos(\theta - \varphi) \quad (3.127)$$

We have $\text{Im}(T_*^{-1}(\xi)) < 0$ and moreover

Lemma 3.10. *Suppose that $\xi \in \Omega_*^-$ and ψ_* is defined by $\xi = \nu_* \cos\psi_*$, $* \in \{L, T, 0\}$. The poles of the matrix function $\varsigma \mapsto TM(\xi, \varsigma)$ located in the half-plane $\text{Im}\,\xi < 0$ are the points $T_*^{-1}(\xi) = \nu_* \cos(\psi_* - \varphi)$ with residue*

$$\text{Res}_\zeta\left(TM(\xi,\zeta),T_*^{-1}(\xi)\right) = -\frac{1}{2i\pi}\,\frac{\sin(\psi_* - \varphi)}{\sin\psi_*}tm_*(T_*^{-1}(\xi)). \qquad (3.128)$$

Proof. Let $\xi \in \mathbb{C}$, $\text{Im}\,\xi < 0$ and $\psi_* = \psi_*^1 + i\psi_*^2 \in \mathcal{D}$ the complex angle such that $\xi = \nu_* \cos\psi_*$. The poles of $\zeta \mapsto TM(\xi,\zeta)$ are those of the functions $\zeta \mapsto D_*(\xi,\zeta)$, $* \in \{L,T,0\}$. If $\zeta = \nu_* \cos\theta$, we have

$$D_*(\xi,\zeta) = \frac{1}{\xi - T_*(\zeta)} = \frac{1}{\nu_*(\cos\psi_* - \cos(\theta + \varphi))}. \qquad (3.129)$$

Hence, the poles of $D_*(\xi,\zeta)$ located in the half-plane $\text{Im}\,\zeta < 0$ are $\zeta = \nu_* \cos\theta$, $\text{Im}\,\theta > 0$ with θ solution of the equation

$$\cos(\theta + \varphi) = \cos\psi_*. \qquad (3.130)$$

(3.130) has at most one solution which is $\theta_* = \psi_* - \varphi$ if $\varphi < \psi_*^1 < \pi$. Hence $D_*(\xi,\zeta)$ has exactly one pole in $\text{Im}\,\zeta < 0$ which is $\zeta = \nu_* \cos\theta_* = T_*^{-1}(\xi)$ if $\xi \in \Omega_*^-$. The residue (3.128) is obtained by taking the limit when $\theta \to \psi_* - \varphi$ of $D_*(\xi,\zeta)(\zeta - T_*^{-1}(\xi))$. ∎

Let W be the domain

$$W = \{\xi, \text{Im}\,\xi < 0, \xi \notin \cup\partial\Omega_*^-\}. \qquad (3.131)$$

We define the *transfer matrix* with index $* \in \{L,T,0\}$ by

$$\begin{cases} M_*(\xi) = -\dfrac{\sin\chi_*}{\sin\psi_*}dm(\xi)^{-1}tm_*(T_*^{-1}(\xi)) \\ \xi = \nu_* \cos\psi_*, \chi_* = \psi_* - \varphi, T_*^{-1}(\xi) = \nu_* \cos\chi_*. \end{cases} \qquad (3.132)$$

The deformation of the contour Γ_0 onto the contour Γ_2 yields now

$$\begin{aligned} \int_{\Gamma_0} TM(\xi,\zeta)\cdot f(\zeta)d\zeta = &\int_{\Gamma_2} TM(\xi,\zeta)\cdot f(\zeta)d\zeta \\ &- \sum_{*\in\{L,T,0\}} 1(\xi \in \Omega_*^-)M_*(\xi)\cdot f(T_*^{-1}(\xi)) \end{aligned} \qquad (3.133)$$

Defining the functions g_1, g_2, holomorphic in W, by

$$\begin{cases} g_1(\xi) = dm(\xi)^{-1}\left[\dfrac{W_1}{\xi - Z_1}\right. \\ \qquad \left. - \int_{\Gamma_2} DM(\xi,\zeta)\cdot \Sigma_1(\zeta)d\zeta - \int_{\Gamma_2} TM(\xi,\zeta)\cdot \Sigma_2(\zeta)d\zeta\right] \\ g_2(\xi) = dm(\xi)^{-1}\left[\dfrac{W_2}{\xi - Z_2}\right. \\ \qquad \left. - \int_{\Gamma_2} TM(\xi,\zeta)\cdot \Sigma_1(\zeta)d\zeta - \int_{\Gamma_2} DM(\xi,\zeta)\cdot \Sigma_2(\zeta)d\zeta\right] \end{cases} \qquad (3.134)$$

We obtain the following result

Lemma 3.11 (Functional equation for the spectral function). *For $\xi \in W$, the spectral function is solution of the functional equation*

$$
\begin{cases}
\Sigma_1(\xi) = g_1(\xi) + \displaystyle\sum_{*\in\{L,T,0\}} M_*(\xi) \cdot \Sigma_2(T_*^{-1}(\xi))\mathbf{1}(\xi \in \Omega_*^-) \\[4mm]
\Sigma_2(\xi) = g_2(\xi) + \displaystyle\sum_{*\in\{L,T,0\}} M_*(\xi) \cdot \Sigma_1(T_*^{-1}(\xi))\mathbf{1}(\xi \in \Omega_*^-)
\end{cases}
\tag{3.135}
$$

where the transfer matrices $M_*(\xi)$ and the scalar $T_*^{-1}(\xi)$ are given by

- $M_L(\xi) = -\dfrac{1}{\delta(\xi)}\Big[\cos\chi z_L(\xi); -\sin\chi z_L(\xi); 0\Big]$ (3.136)

  $T_L^{-1}(\xi) = \nu_L \cos\chi$, *where* $\xi = \nu_L \cos\psi_L, \chi = \psi_L - \varphi$

- $M_T(\xi) = -\dfrac{1}{\delta(\xi)}\Big[\sin\chi z_T(\xi); \cos\chi z_T(\xi); 0\Big]$ (3.137)

  $T_T^{-1}(\xi) = \nu_T \cos\chi$, *where* $\xi = \nu_T \cos\psi_T, \chi = \psi_T - \varphi$

- $M_0(\xi) = \begin{bmatrix} 0 & 0 & 0 \\ 0 & 0 & 0 \\ 0 & 0 & -1 \end{bmatrix}$ (3.138)

  $T_0^{-1}(\xi) = \nu_0 \cos\chi$, *where* $\xi = \nu_0 \cos\psi_0, \chi = \psi_0 - \varphi$

(see (3.116-3.118) for $\delta(\xi), z_L(\xi), z_T(\xi)$).

Proof. The formulas for the transfer matrices result from Lemma 3.10 and of (3.117-3.118). ∎

4. Proofs of the Results

4.1 An Isomorphism Theorem

In this section, we prove an isomorphism theorem which is the cornerstone of the study. Recall that $\mathbb{U}$ is the domain (cf Sect.3.3)

$$\mathbb{U} = \mathbb{C} \backslash] -\infty, -1], \tag{4.1}$$

and that $\mathcal{H}$ is the functional space

$$\mathcal{H} = \{ f \in Hol(\mathbb{U}) \ / \ f(xe^{i\theta}) \in H^+, \ \forall \, \theta \in]0, \pi[\} . \tag{4.2}$$

$Hol(\mathbb{U})$ is the space of the functions holomorphic on $\mathbb{U}$, and we refer to Sect.3.3 for the definition of H^+. The integral kernels DM^θ, TM^θ are given, for $\theta \in [0, \pi[$, by the relations (3.6), (3.9), or equivalently by

$$\begin{cases} DM^\theta(\xi, \zeta) = e^{i\theta} DM(\xi e^{i\theta}, \zeta e^{i\theta}) \\ TM^\theta(\xi, \zeta) = e^{i\theta} TM(\xi e^{i\theta}, \zeta e^{i\theta}). \end{cases} \tag{4.3}$$

We call $\mathcal{L}_\theta$ the 6×6 system of integral equations

$$\mathcal{L}_\theta \begin{bmatrix} X_1 \\ X_2 \end{bmatrix} = \begin{pmatrix} DM^\theta & TM^\theta \\ TM^\theta & DM^\theta \end{pmatrix} \begin{bmatrix} X_1 \\ X_2 \end{bmatrix} = \begin{bmatrix} Y_1 \\ Y_2 \end{bmatrix}. \tag{4.4}$$

Lemmas 3.6, 3.7 state that the operator $\mathcal{L}_\theta$ is bounded on $(H^+)^3 \oplus (H^+)^3$. The main result of this paragraph is the following isomorphism theorem

Theorem 4.1. *For* $\theta \in]0, \pi[$, $\mathcal{L}_\theta$ *is an isomorphism on* $(H^+)^3 \oplus (H^+)^3$.

Before proving this result, we deduce the following corollary.

Proposition 4.2. *Let* $Y = \begin{bmatrix} Y_1 \\ Y_2 \end{bmatrix} \in \mathcal{H}^3 \oplus \mathcal{H}^3$. *In particular,* $Y(ix) \in (H^+)^3 \oplus (H^+)^3$. *If* $X[ix]$ *is the solution of* $\mathcal{L}_{\pi/2} X(ix) = Y(ix)$, *then* $X \in \mathcal{H}^3 \oplus \mathcal{H}^3$.

Proof. For $\theta \in]0, \pi[$, we have $Y(e^{i\theta}x) \in (H^+)^3$. We call $X^\theta(e^{i\theta}x)$ the solution of $\mathcal{L}_\theta X^\theta(e^{i\theta}x) = Y(e^{i\theta}x)$ given by Theorem 4.1.

To begin, we check that the functions X^θ define without ambiguity an holomorphic function $X(z)$ on the domain $\mathbb{C}\backslash]-\infty, 0]$. By construction of $\mathcal{L}_\theta$, it is sufficient to check that the function $X^\theta(e^{i\theta}x)$, which lies in $(H^+)^3$, can be extended analytically to the angular domain $\Gamma_{\alpha_0} = \{x \in \mathbb{C}, \, /|\text{Arg } x| \leq \alpha_0\}$ for a small $\alpha_0 > 0$ and is such that $X^\theta\left(e^{i(\theta+\alpha)}x\right) \in (H^+)^3$ for $|\alpha| \leq \alpha_0$. We have

$$DM^\theta(X_1^\theta) = Y_1(e^{i\theta}x) - TM^\theta(X_2^\theta) = Z(x). \tag{4.5}$$

It follows from Lemma 3.7 that $Z(xe^{i\alpha}) \in (H^+)^3$ for small values of $|\alpha|$. Moreover, we have

$$DM^\theta(\xi, \zeta) = \frac{1}{2i\pi} \frac{1}{\xi - \zeta} dm(\zeta e^{i\theta}). \tag{4.6}$$

The properties of the Hilbert projector (cf Appendix) imply that the function $x \mapsto dm(xe^{i\theta})$. $X_1(e^{i\theta}x)$ is holomorphic in Γ_{α_0} and is in the space $L^2(xe^{i\alpha})$, $|\alpha| \leq \alpha_0$. The conclusion follows now from Lemma 3.5.

Finally, we must check that the function $X(z)$, which is holomorphic in $\mathbb{C}\backslash] - \infty, 0]$, can be extended to $\mathbb{C}\backslash] - \infty, -1]$. Again we start from

$$\begin{cases} DM^{\pi/2}(X_1(i\zeta)) = Y_1(i\zeta) - TM^{\pi/2}(X_2(i\zeta)) \\ DM^{\pi/2}(X_2(i\zeta)) = Y_2(i\zeta) - TM^{\pi/2}(X_1(i\zeta)) \end{cases} \tag{4.7}$$

and we use

$$D_{L,T,0}^{\pi/2}(\xi, \zeta) = \frac{1}{\xi - \left(\cos\varphi\zeta + i\sin\varphi\sqrt{\nu_{L,T,0}^2 + \zeta^2}\right)}. \tag{4.8}$$

By construction, we have

$$TM^{\pi/2}(X_2(i\zeta))\left(\frac{\xi}{i}\right) = \int_\Gamma TM(\xi, \zeta)X_2(\zeta)d\zeta = \Theta(\xi), \tag{4.9}$$

where Γ is the imaginary axis with orientation from bottom to top. Using the conformal representation introduced in Sect.2.4, we obtain that $\Theta(\xi)$ extends to $\mathbb{C}\backslash] - \infty, -1]$ if $\varphi \geq \frac{\pi}{2}$ and to the right of the hyperbola $\{z = \cos\left(\frac{\pi}{2} + \varphi + it\right), \, t \in \mathbb{R}\}$ if $\varphi < \frac{\pi}{2}$. Consequently, the right-hand side in (4.7) extends in a similar way, hence $X(\xi)$ too, by using Lemma 3.5 and the properties of the Hilbert projector. Iterating a finite number of times this argument on equation (4.7), we get Proposition 4.2. ∎

Proof of Theorem 4.1. Let $\theta \in]0,\pi[$ be fixed. For $\gamma \in]0,\pi[\cup]\pi,2\pi[$, we note $\Omega_\pm$ the two domains (Fig. 4.1)

$$\begin{cases} \Omega_+ = \{x = r\cos\varphi,\ y = r\sin\varphi,\ \varphi \in]0,\gamma[\} \\ \Omega_- = \{x = r\cos\varphi,\ y = r\sin\varphi,\ \varphi \in]\gamma,2\pi[\} \end{cases} \qquad (4.10)$$

and $\Gamma = \partial\Omega_+ = \partial\Omega_-$ their common boundary.

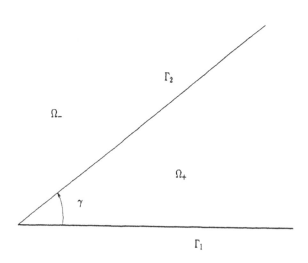

Fig. 4.1. The domains Ω_+ and Ω_-.

We call $\Gamma = \Gamma^1 \cup \Gamma^2$, $\Gamma^1 = \{\varphi = 0\}$, $\Gamma^2 = \{\varphi = \gamma\}$. We note E^+ (resp. E^-) the space of the distributions f on Ω_+ (resp. Ω_-) satisfying in polar coordinates

$$f \in C^0\left(I, H_r^1(\mathbb{R}_+)\right)\ ;\ \frac{1}{r}\partial_\varphi f \in C^0\left(I, L^2(\mathbb{R}_+)\right) \qquad (4.11)$$

where $I = [0,\gamma]$ (resp. $I = [\gamma,2\pi]$).

For $f \in E^\pm$, we have $f_{|\Gamma} \in H^1(\Gamma)$, because $f_{|r=0}$ is independent of φ. Indeed, we have $f(r,\varphi_1) - f(r,\varphi_2) = r\int_{\varphi_1}^{\varphi_2} \frac{1}{r}\partial_\varphi f(\sigma,r)d\sigma \in rL^2(\mathbb{R}_+) \cap H^1(\mathbb{R}_+)$. We define the four operators

$$\begin{cases} A_+(u) = -(\lambda + \mu)\text{grad div } u - \mu \Delta u - e^{-2i\theta}u & \text{elasticity operator} \\ A_-(g) = -\Delta g - \nu_0^2 e^{-2i\theta}g & \text{wave operator} \\ B_+(u) = (\lambda \text{ div } u + 2\mu\varepsilon(u)).\mathbf{n}_{|\Gamma} & \text{normal stress} \\ B_-(g) = \nabla g.\mathbf{n}_{|\Gamma} & \text{normal velocity} \end{cases}$$

$$(4.12)$$

and the hermitian bilinear forms

$$\begin{cases} \sigma_+(u,v) = \displaystyle\int_{\Omega_+} \lambda \text{ div } u.\text{div } \bar{v} + 2\mu \text{ tr}(\varepsilon(u)\varepsilon(\bar{v})) - e^{-2i\theta}u\bar{v}, \\ \sigma_-(u,v) = \displaystyle\int_{\Omega_-} \nabla g \nabla \bar{f} - \nu_0^2 e^{-2i\theta}g\bar{f}. \end{cases}$$

$$(4.13)$$

The two forms $\sigma_\pm$ are coercive on the Sobolev space H^1 (for σ_+, this is the Korn inequality). Moreover the integration by parts formulas are

$$\begin{cases} \displaystyle\int_{\Omega_+} A_+(u)\bar{v} = \sigma_+(u,v) + \int_\Gamma B_+(u)\bar{v}, \\ \displaystyle\int_{\Omega_-} A_-(g)\bar{f} = \sigma_-(g,f) - \int_\Gamma B_-(g)\bar{f}. \end{cases}$$

$$(4.14)$$

We have now the following existence and uniqueness lemmas.

Lemma 4.3 (coupled system). *For* $w \in L^2(\Omega_+)^2$, $h \in L^2(\Omega_-)$, $k \in L^2(\Gamma)^2$, $\ell \in L^2(\Gamma)$, *the coupled system*

$$\begin{cases} A_-(g) = h & \text{in } \Omega_- \\ A_+(u) = w & \text{in } \Omega_+ \\ B_+(u) - ie^{-i\theta}\rho g\mathbf{n} = k & \text{on } \Gamma \\ ie^{-i\theta}\mathbf{u}.\mathbf{n} - \text{grad } g.\mathbf{n} = \ell & \text{on } \Gamma \end{cases}$$

$$(4.15)$$

does have an unique solution (u,g) *in the space* $u \in (H^1(\Omega_+) \cap E^+)^2$, $g \in (H^1(\Omega_-) \cap E^-)$.

Lemma 4.4 (elasticity problem). *The Neumann elasticity problem*

$$\begin{cases} A_+(u) = w \in L^2(\Omega_+)^2 \\ B_+(u) = k \in L^2(\Gamma)^2 \end{cases}$$

$$(4.16)$$

and the Dirichlet elasticity problem

$$\begin{cases} A_+(u) = w \in L^2(\Omega_+)^2 \\ u_{|\Gamma} = k \in H^1(\Gamma)^2 \end{cases}$$

$$(4.17)$$

do have an unique solution $u \in (H^1(\Omega_+) \cap E^+)^2$.

Lemma 4.5 (wave problem). *The Neumann wave problem*

$$\begin{cases} A_-(g) = h \in L^2(\Omega_-) \\ B_-(g) = \ell \in L^2(\Gamma) \end{cases} \tag{4.18}$$

and the Dirichlet wave problem

$$\begin{cases} A_-(g) = h \in L^2(\Omega_-) \\ g_{|\Gamma} = \ell \in H^1(\Gamma) \end{cases} \tag{4.19}$$

do have an unique solution $g \in H^1(\Omega_-) \cap E^-$.

Let us now check that the three preceeding lemmas imply the Theorem 4.1.

• *Injectivity of $\mathcal{L}_\theta$:*

If $X_1(\xi) = \left(\hat{\alpha}_1(\xi), \hat{\beta}_1(\xi), \hat{\gamma}_1(\xi) \right)$, $X_2(\xi) = \left(\hat{\alpha}_2(\xi), \hat{\beta}_2(\xi), \hat{\gamma}_2(\xi) \right)$ are solutions of

$$\mathcal{L}_\theta \begin{bmatrix} X_1 \\ X_2 \end{bmatrix} = 0. \tag{4.20}$$

We define

$$\begin{cases} \alpha = \alpha_1(r)\delta_{\Gamma_1} + \alpha_2(r)\delta_{\Gamma_2} \\ \beta = \beta_1(r)\delta_{\Gamma_1} + \beta_2(r)\delta_{\Gamma_2} \\ \gamma = \gamma_1(r)\delta_{\Gamma_1} + \gamma_2(r)\delta_{\Gamma_2} \end{cases} \tag{4.21}$$

where $\alpha_j(r)$, $\beta_j(r)$, $\gamma_j(r)$ are the inverse Fourier transforms of $\hat{\alpha}_j(\xi), \hat{\beta}_j(\xi), \hat{\gamma}_j(\xi)$ and δ_{Γ_j} the distribution defined by

$$\langle \delta_{\Gamma_j}, \varphi \rangle = \int_{\Gamma_j} \varphi. \tag{4.22}$$

Let u, g be the tempered distributions on $\mathbb{R}^2$ defined by

$$A_+(u) = \begin{bmatrix} \alpha \\ \beta \end{bmatrix} ; \quad A_-(g) = \gamma. \tag{4.23}$$

We have $\alpha_i, \beta_i, \gamma_i \in L^2(\mathbb{R}_+)$. Therefore $\alpha, \beta, \gamma \in H^{-\frac{1}{2}-\varepsilon}(\mathbb{R}^2)$ for each $\varepsilon > 0$ and $(u, g) \in H^{\frac{3}{2}-\varepsilon}(\mathbb{R}^2)$. Moreover (cf. [L1], Sect.2.3, Prop. 3), we have

$$u_{|\Omega_\pm} \in (E_\pm)^2 ; \quad g_{|\Omega_\pm} \in E_\pm. \tag{4.24}$$

We deduce from the construction of Sect.2,Sect.3 that $\mathcal{L}_\theta \begin{bmatrix} X_1 \\ X_2 \end{bmatrix} = 0$ implies that (u, g) is solution of (4.15) with zero right-hand side. Lemma 4.3 implies now $(u, g) = (0, 0)$, therefore $X_1 = X_2 = 0$.

- *Surjectivity of $\mathcal{L}_\theta$* : For $Y = (Y_1, Y_2) \in (H^+)^3 \oplus (H^+)^3$, we set

$$Y_j(\xi) = \frac{1}{2} \begin{pmatrix} \hat{a}_j(\xi) \\ \hat{b}_j(\xi) \\ \hat{c}_j(\xi) \end{pmatrix}, \quad j = 1, 2 \tag{4.25}$$

where $a_j(r), b_j(r), c_j(r) \in L^2(\mathbb{R}_+)$, and we define $k = (k_1, k_2) \in L^2(\Gamma)^2$, $\ell \in L^2(\Gamma)$ by

$$k_1|_{\Gamma_j} = a_j; \quad k_2|_{\Gamma_j} = b_j; \quad \ell|_{\Gamma_j} = c_j ; \quad j = 1, 2. \tag{4.26}$$

Let (u_+, g_-) be the solution of (4.15) with right-hand side $h = 0$, $w = 0$, k, ℓ. Let (u_-, g_+) be the solution given by Lemmas 4.4, 4.5 (Dirichlet case) of the problems

$$\begin{cases} u_- \in H^1(\Omega_-)^2 \cap E_-^2 \\ A_+(u_-) = 0 \quad \text{in } \Omega_- \\ u_-|_\Gamma = u_+|_\Gamma \in H^1(\Gamma)^2, \end{cases} \tag{4.27}$$

and

$$\begin{cases} g_+ \in H^1(\Omega_+)^2 \cap E_+ \\ A_-(g_+) = 0 \quad \text{in } \Omega_+ \\ g_+|_\Gamma = g_-|_\Gamma \in H^1(\Gamma). \end{cases} \tag{4.28}$$

The functions u, g defined on $\mathbb{R}^2$ by

$$u|_{\Omega_\pm} = u_\pm ; \quad g|_{\Omega_\pm} = g_\pm \tag{4.29}$$

are such that $u \in H^1(\mathbb{R}^2)^2$, $g \in H^1(\mathbb{R}^2)$ and are solutions in $\mathcal{D}'(\mathbb{R}^2)$ of the equations

$$\begin{cases} A_+(u) = -[B_+(u_+) - B_+(u_-)] \otimes \delta_\Gamma = \begin{pmatrix} \alpha \\ \beta \end{pmatrix} \otimes \delta_\Gamma \\ A_-(g) = -[B_-(g_+) - B_-(g_-)] \otimes \delta_\Gamma = \gamma \otimes \delta_\Gamma \end{cases} \tag{4.30}$$

Because of the jump formula, we check that the equations (4.30) are satisfied at $(x, y) \neq (0, 0)$. In addition, $A_+(u) \in H^{-1}(\mathbb{R}^2)^2$, $A_-(g) \in H^{-1}(\mathbb{R}^2)$ and there is no nontrivial distribution in $H^{-1}(\mathbb{R}^2)$ with support $(0, 0)$.

To conclude, the functions $\alpha_j = \alpha_{|\Gamma_j}$, $\beta_j = \beta_{|\Gamma_j}$, $\gamma_j = \gamma_{|\Gamma_j}$, $j = 1, 2$, are in $L^2(\mathbb{R}_+)$ and if $X_j = \left(\hat{\alpha}_j(\xi), \hat{\beta}_j(\xi), \hat{\gamma}_j(\xi) \right)$ we have by the construction of Sect.2

$$\mathcal{L}_\theta(X) = Y. \tag{4.31}$$

It remains now to prove Lemmas 4.3, 4.4, 4.5 .

Proof of Lemma 4.5. Lemma 4.5 is proved in [L1], Sect.4.2, Lemma 2.

Proof of Lemma 4.3. Since the integrations by parts hold in the space $H^1 \cap E$, the system (4.15) is equivalent to the weak problem: look for $(u, g) \in H^1(\Omega_+)^2 \oplus H^1(\Omega_-)$ such that for each $(v, f) \in H^1(\Omega_+)^2 \oplus H^1(\Omega_-)$

$$\Sigma\{(u, g), (v, f)\} = \int_{\Omega_+} w\bar{v} + \int_{\Omega_-} \rho h \bar{f} - \int_\Gamma \left[k\bar{v} + \rho \ell \bar{f} \right], \tag{4.32}$$

where Σ is the bilinear form defined by

$$\Sigma\{(u, g), (v, f)\} = \sigma_+(u, v) + \rho\sigma_-(g, f) + ie^{-i\theta}\rho \int_\Gamma g(\mathbf{n}.\bar{v}) - (u.\mathbf{n})\bar{v}. \tag{4.33}$$

The coercivity of the form Σ results from

$$\begin{cases} \mathrm{Im}\left(e^{i\theta} \Sigma\{(u, g), (u, g)\} \right) = \sin\theta \left(\sigma_+^0 + \sigma_-^0 \right) \\ \sigma_+^0 = \int_{\Omega_+} \lambda |\mathrm{div}\, u|^2 + 2\mu\, \mathrm{tr}(\varepsilon(u), \varepsilon^*(u)) + |u|^2 \\ \sigma_-^0 = \int_{\Omega_-} |\nabla g|^2 + \nu_0^2 |g|^2. \end{cases} \tag{4.34}$$

Thus, we obtain the existence and uniqueness of the solution (u, g) of the problem (4.32) in the space H^1. It remains to check the regularity, that is, $u \in (E^+)^2$, and $g \in E^-$. We remark that the coupling term occuring in Σ

$$ie^{-i\theta}\rho \int_\Gamma g\mathbf{n}.\bar{v} - u.\mathbf{n}.\bar{f} \tag{4.35}$$

can be put in the right-hand side, because $g_{|\Gamma}, u_{|\Gamma}$ are in $L^2(\Gamma)$. This is equivalent to the transformation $k \to k + ie^{-i\theta}\rho g\mathbf{n}$, $\ell \to -ie^{-i\theta}\mathbf{u}.\mathbf{n}$ on the initial problem.

Now, the problem is decoupled into a problem (4.16) for u and a problem (4.18) for g. Consequently, Lemma 4.3 results from Lemmas 4.4 and 4.5.

∎

Proof of Lemma 4.4. The variational form of the problem (4.16) reads :

- look for $u \in H^1(\Omega_+)^2$ such that for each $v \in H^1(\Omega_+)^2$

$$\sigma_+(u, v) = \int_{\Omega_+} w\bar{v} - \int_\Gamma k\bar{v}. \tag{4.36}$$

For the problem (4.17), we have

- look for $u \in H^1(\Omega_+)^2$ such that for $v \in H^1_0(\Omega_+)^2$

$$\begin{cases} \sigma_+(u, v) = \int_\Omega w\bar{v} \\ u_{|\Gamma} = k \in H^1(\Gamma). \end{cases} \tag{4.37}$$

We have the existence and uniqueness of $u \in H^1(\Omega_+)^2$, solution of the problems (4.36) or (4.37). For (4.37), if $u_1 \in H^1(\Omega_+)$, $u_{1|\Gamma} = k$, is a lifting of k, we have that $u = u_1 + u_2$, satisfies

$$\sigma_+(u_2, v) = -\sigma_+(u_1, v) + \int_\Omega u\bar{v} \quad \forall\, v \in H^1_0. \tag{4.38}$$

It remains to check the regularity of $u \in (E_+)^2$ for the problems (4.36), (4.37). We begin by localizing the problem in a neighborhood of the vertex of the wedge ($r = 0$) by writing $1 = \chi_0 + \chi_1 + \chi_2 + \chi_3$ where $\chi_j \in C^\infty(\mathbb{R}^2)$, are localization functions. The function χ_0 localizes in the ball $r \leq 2$, and χ_1 (resp. χ_2) in a neighborhood of $\{\Gamma_1 \cap r \geq 1\}$ (resp. $\{\Gamma_2 \cap r \geq 1\}$), (Fig. 4.2). The functions χ_j are C^∞, with bounded derivatives and such that $\delta_n\chi_{j|\Gamma} = 0$.

- *Regularity of u_3*

    The support of $u_3 = \chi_3 u$ does not intersect Γ. We have

$$A_+(u_3) = \chi_3 w + [A_+, \chi_3]u_3 \in L^2(\mathbb{R}^2)^2 \tag{4.39}$$

(where $[A, B] = AB - BA$). Therefore, $u_3 \in H^2(\mathbb{R}^2)^2 \subset (E^+)^2$.

- *Regularity of u_1, u_2*

    We limit ourselves to check the regularity of $u_1 = \chi_1 u$ for the Neumann problem. We have

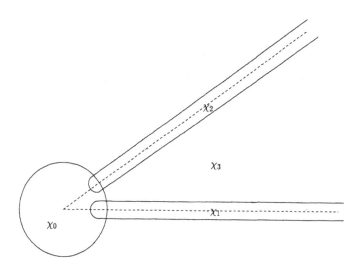

Fig. 4.2. The partition of unity of the wedge.

$$\begin{cases} A_+(u_1) \in L^2 \left(\{(x,y) \in \mathbb{R}^2,\ y > 0\} \right) \\ B_+(u_1) \in L^2(\mathbb{R}) \end{cases} \tag{4.40}$$

with $\mathrm{Supp}(u_1) \subset \{(x = r \cos \varphi, y = r \sin \varphi),\ r \geq \frac{1}{2},\ 0 \leq \varphi < \gamma_0,\ \gamma_0 \text{ small}\}$. Up to a subtraction from u_1 of a function in H^2, the problem is equivalent to check that the solution of

$$\begin{cases} A_+(u) = 0 \ \text{ in } y > 0 \\ (\lambda \ \mathrm{div}\ u + 2\mu\varepsilon(u)).\mathbf{n}_{|y=0} = k(x) \in L^2(\mathbb{R}) \end{cases} \tag{4.41}$$

lies in the space

$$u \in C^0(J, H^1_r(\mathbb{R}_+)) \cap C^1(J, L^2(\mathbb{R}_+)), \tag{4.42}$$

where $J = [0, \varphi_0]$ with φ_0 small. The explicit computation of the solution of (4.41) yields

$$u(x, y) = \int e^{i(x\xi + y\eta)} M_\theta^{-1} \begin{pmatrix} a(\xi) \\ b(\xi) \end{pmatrix} d\xi d\eta, \tag{4.43}$$

where M_θ is the symbol matrix of the operator A_+

$$M_\theta = \begin{pmatrix} (\lambda + \mu)\xi^2 + \mu(\xi^2 + \eta^2) - e^{-2i\theta} & (\lambda + \mu)\xi\eta \\ (\lambda + \mu)\xi\eta & (\lambda + \mu)\eta^2 + \mu(\xi^2 + \eta^2) - e^{-2i\theta} \end{pmatrix} \tag{4.44}$$

and where $a, b \in L^2$. The result follows now from the proof of Proposition 3 of [L1], Sect.2.3.

• *Regularity of u_0*

We suppose now that in (4.36), (4.37), the functions u, w, k, h have their support in $r \leq 2$. We put the term $e^{-2i\theta} u\bar{v}$ in the right-hand side. In other words, we pose

$$\sigma(u, v) = \int_\Omega \lambda \operatorname{div} u \operatorname{div} v + 2\mu \operatorname{tr}(\varepsilon(u)\varepsilon(v)) \tag{4.45}$$

and we consider the two problems : look for $u \in (H^1)^2$ such that

$$\sigma(u, v) = \int_\Omega wv - \int_\Gamma kv \quad \forall\, v \in H^1 \quad \text{(Neumann)} \tag{4.46}$$

$$\begin{cases} \sigma(u, v) = \int_\Omega wv & \forall\, v \in H_0^1 \quad \text{(Dirichlet)} \\ u_{|\Gamma} = h. \end{cases} \tag{4.47}$$

In order to study the regularity of u in (4.46), (4.47), we use the Mellin transform, defined by

$$f(r) \to (Mf)(z) = \int_0^{+\infty} f(r) r^{-z} \frac{dr}{r} = F(z), \tag{4.48}$$

which is an isometry (up to the factor 2π) from $L^2(\mathbb{R}_+)$ onto $L^2 \left(\operatorname{Re} z = -\frac{1}{2} \right)$. Its inverse is

$$(M^{-1}F)(r) = \frac{1}{2i\pi} \int_{\operatorname{Re}\, z = -\frac{1}{2}} F(z) r^z dz. \tag{4.49}$$

Moreover, we write the vector field u in polar coordinates in the form

$$u = u_r \begin{pmatrix} \cos \varphi \\ \sin \varphi \end{pmatrix} + u_\varphi \begin{pmatrix} -\sin \varphi \\ \cos \varphi \end{pmatrix}. \tag{4.50}$$

We note

$$a = \lambda + 2\mu, \quad b = \mu \tag{4.51}$$

and

$$(Mf)(z) = \hat{f}(z). \tag{4.52}$$

If $A_+^0 = -(\lambda + \mu)\text{grad div} - \mu\Delta$ is the elasticity operator, the operator $-r^2 A_+^0$ reads in polar coordinates

$$-r^2 A_+^0 = \begin{bmatrix} a\left[(r\partial_r)^2 - 1\right] + b\partial^2\varphi & \left[(a - b)r\partial_r - (a + b)\right]\partial_\varphi \\ (a - b)r\partial_r + (a + b)\partial_\varphi & b\left[(r\partial_r)^2 - 1\right] + a\partial^2\varphi \end{bmatrix} ; \tag{4.53}$$

hence

$$M(-r^2 A_+^0(u))(z) = -A_z \begin{bmatrix} \hat{u}_r(z, \varphi) \\ \hat{u}_\varphi(z, \varphi) \end{bmatrix} \tag{4.54}$$

where

$$-A_z = \begin{bmatrix} b\partial^2\varphi + a(z^2 - 1) & \left[(a - b)z - (a + b)\partial_\varphi\right] \\ \left[(a - b)z + (a + b)\right]\partial_\varphi & a\partial^2\varphi + b(z^2 - 1) \end{bmatrix} . \tag{4.55}$$

Similarly we have

$$M(rB_+(u)(z)) = B_z \begin{bmatrix} \hat{u}_r \\ \hat{u}_\varphi \end{bmatrix} \tag{4.56}$$

where

$$B_z = \begin{bmatrix} b\partial_\varphi & b(z - 1) \\ \lambda z + a & a\partial_\varphi \end{bmatrix} . \tag{4.57}$$

We use now in (4.36) or (4.37) the trial vector fields v given in polar coordinates by

$$v = r^{-z} \left[v_r(\varphi) \begin{pmatrix} \cos\varphi \\ \sin\varphi \end{pmatrix} + v_\varphi(\varphi) \begin{pmatrix} -\sin\varphi \\ \cos\varphi \end{pmatrix} \right] = r^{-z}\tilde{v}(\varphi) \tag{4.58}$$

with $\text{Re } z < 0$ and $v_r, v_\varphi \in H^1(0, \gamma)$ (or $H_0^1(0, \gamma)$ in the Dirichlet case). Since the supports of u, w, k, h are included in the ball $r \leq 2$, the identities (4.36), (4.37) hold for each vector field v of the form (4.58).

We have

$$\int_\Omega w.v = \int_0^\gamma \hat{w}(z - 2, \varphi).\tilde{v}(\varphi)d\varphi, \tag{4.59}$$

$$\int_{\Gamma_j} k.v = \hat{k}_j(z - 1).\tilde{v}(\varphi_j) \quad (\varphi_1 = 0, \ \varphi_2 = \gamma) \tag{4.60}$$

and

$$\sigma(u, v) = \int_0^\gamma q_z(\hat{u}, \tilde{v})d\varphi \tag{4.61}$$

with

$$q_z(\hat{u}, \tilde{v}) = a(\partial_\varphi \hat{u}_\varphi + \hat{u}_r)(\partial_\varphi \tilde{v}_\varphi + \tilde{v}_r)$$
$$+ b\left[\partial_\varphi \hat{u}_r - \hat{u}_\varphi\right]\left[\partial_\varphi \tilde{v}_r - \tilde{v}_\varphi\right] - bz^2 \hat{u}_\varphi \tilde{v}_\varphi$$
$$- az^2 \hat{u}_r \tilde{v}_r + \lambda z \left[\hat{u}_r \partial_\varphi \tilde{v}_\varphi - \tilde{v}_r \partial_\varphi \hat{u}_\varphi\right]$$
$$+ \mu z \left[\hat{u}_\varphi \partial_\varphi \tilde{v}_r - \tilde{v}_\varphi \partial_\varphi \hat{u}_r\right]. \tag{4.62}$$

The integration by parts formula reads

$$\int_0^\gamma A_z(\hat{u})\tilde{v}\,d\varphi = \int_0^\gamma q_z\,(\hat{u}, \tilde{v})\,d\varphi + [B_z(\hat{u}).\tilde{v}]_0^\gamma. \tag{4.63}$$

Let us introduce for $z \in \mathbb{C}$ the Hilbert spaces L_z^2, H_z^1, H_z^2 of functions $f(\varphi)$ defined by the norms

$$(4.64)\quad L_z^2: \qquad \int_0^\gamma |f|^2$$

$$(4.65)\quad H_z^1: \qquad \int_0^\gamma |\partial_\varphi f|^2 + (1 + |z|^2)\,|f|^2$$

$$(4.66)\quad H_z^2: \qquad \int_0^\gamma |\partial_\varphi^2 f|^2 + (1 + |z|^2)\,|\partial_\varphi f|^2 + (1 + |z|^4)\,|f|^2$$

and $\mathbb{C}_{z,i}$ the trace spaces

$$\mathbb{C}_{z,i} = \mathbb{C}\,;\quad \|a; \mathbb{C}_{z,i}\|^2 = (1 + |z|)^{1+2(1-i)}|a|^2. \tag{4.67}$$

In addition, we consider the strip of the complex plane defined for α_0 small by

$$U = \left\{z;\ \mathrm{Re}\ z \in \left[-\frac{1}{2} - \alpha_0, \frac{1}{2} + \alpha_0\right]\right\} \tag{4.68}$$

and for $\varepsilon > 0$, $U_\varepsilon = U \cap \{|z| > \varepsilon\}$ (Fig. 4.3). Finally, we consider the operators T_z^N, T_z^D defined by

$$\begin{cases} T_z^N: \ (\mathcal{H}_z^2)^2 \to (L_z^2)^2 \oplus \mathbb{C}_{z,1}^4 \\ \qquad g \mapsto (A_z(g); B_z(g)_{\varphi=0},\ B_z(g)_{\varphi=\gamma}) \end{cases} \tag{4.69}$$

$$\begin{cases} T_z^D: \ (\mathcal{H}_z^2)^2 \to (L_z^2)^2 \oplus \mathbb{C}_{z,0}^4 \\ \qquad g \mapsto (A_z(g); g_{\varphi=0}, g_{\varphi=\gamma}) \end{cases} \tag{4.70}$$

Lemma 4.4 results from

Proposition 4.3. *For α_0 sufficiently small, $T_z^{N,D}$ is a Fredholm operator with index 0 for $z \in U$, invertible for $z \in U_\varepsilon$ and*

$$\|(T_z^{N,D})^{-1}\| \le C(\varepsilon), \quad z \in U_\varepsilon. \tag{4.71}$$

Let us check first that the Proposition 4.3 allows to conclude the proof of Lemma 4.4.

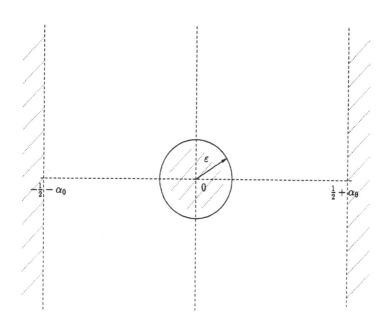

Fig. 4.3. The strip U.

• *Neumann case*

For $z \in U$, Re $z < 0$, (4.36), (4.59), (4.60), (4.61) and (4.63) yield

$$
\begin{cases}
A_z(\hat{u}) = \hat{w}(z - 2, \varphi) \\
B_z(\hat{u})_{\varphi=0} = \hat{k}_1(z - 1) \\
B_z(\hat{u})_{\varphi=\gamma} = \hat{k}_2(z - 1).
\end{cases}
\tag{4.72}
$$

Since $w \in L^2(\Omega_+)$, $k \in L^2(\Gamma)$ have their support in the ball $r \leq 2$, we have

$$\sup_{t<1/2} \int_{\mathrm{Re}\,z=t} \left| \hat{k}(z-1) \right|^2 dz \le C \|k\|_{L^2}, \qquad (4.73)$$

$$\sup_{t<1} \int_0^\gamma \int_{\mathrm{Re}\,z=t} |\hat{w}(z-2,\varphi)|^2 \, dz d\varphi \le C \|w\|_{L^2}. \qquad (4.74)$$

It follows from Proposition 4.3, (4.72), (4.73) and (4.74) that the function $\hat{u}(z,\varphi)$, holomorphic in $\mathrm{Re}\,z < 0$, can be extended to $U_\varepsilon \cap \left(\mathrm{Re}\,z < \frac{1}{2} \right)$ and is such that

$$\sup_{1/4<t<1/2} \int_{\mathrm{Re}\,z=t} dz \int_0^{\gamma_0} d\varphi \left[\frac{|\partial_\varphi^2 \hat{u}|^2}{(1+|z|)} + (1+|z|) \, |\partial_\varphi \hat{u}|^2 + (1+|z|)^3 \, |\hat{u}|^2 \right] < +\infty. \qquad (4.75)$$

Therefore, the trace theorems yield

$$z\hat{u}, \frac{\partial \hat{u}}{\partial \varphi} \in C^0 \left(\varphi \in [0,\gamma_0], L^2 \left(\mathrm{Re}\,z = \frac{1}{2} \right) \right). \qquad (4.76)$$

Using the inverse Mellin transform (4.49) and the Cauchy formula, we get (we note ψ_z the right-hand side in (4.72))

$$\begin{pmatrix} u_r \\ u_\varphi \end{pmatrix} = \frac{1}{2i\pi} \int_{\mathrm{Re}\,z=\frac{1}{2}} \hat{u} r^z \, dz + \frac{1}{2i\pi} \int_{|z|=\varepsilon} \left(T_z^N \right)^{-1} \psi_z dz = I + II. \qquad (4.77)$$

(4.76) gives that $\left(\partial_r, \frac{1}{r}, \frac{1}{r}\partial_\varphi \right)(I) \in C^0 \left(\varphi, L_r^2 \right)$. Writing $I = (I_r, I_\varphi)$, $II = (II_r, II_\varphi)$, we get

$$\begin{pmatrix} I_x \\ I_y \end{pmatrix} = I_r \begin{pmatrix} \cos \varphi \\ \sin \varphi \end{pmatrix} + I_\varphi \begin{pmatrix} -\sin \varphi \\ \cos \varphi \end{pmatrix} ; \qquad (4.78)$$

(4.78) and the analogous formula for II give

$$\begin{pmatrix} u_x \\ u_y \end{pmatrix} = \begin{pmatrix} I_x \\ I_y \end{pmatrix} + \begin{pmatrix} II_x \\ II_y \end{pmatrix} \in H^1. \qquad (4.79)$$

Since $I_x, I_y \in E^+$, the residual terms II_x, II_y belong to H^1 in the neighborhood of $r = 0$. Since these terms are a finite sum of the form

$$II_{x,y} = \sum_{\text{finite}} (\log r)^j f_j(\varphi) \quad j \ge 0 \qquad (4.80)$$

they are constant, therefore $u \in E^+$.   ∎

• *Dirichlet case*

   For $z \in U$, $\mathrm{Re}\,z < 0$, we have

$$\begin{cases} A_z(\hat{u}) = \hat{w}(z - 2, \varphi) \\ \hat{u}_{|\varphi=0} = \hat{h}_1(z) \\ \hat{u}_{|\varphi=\gamma} = \hat{h}_2(z). \end{cases} \qquad (4.81)$$

The functions $\hat{h}_j(z)$ are meromorphic in $\mathrm{Re}\, z < \frac{1}{2}$. The only pole is simple, located at $z = 0$ and its residue is $-h_j(0)$. Moreover, we have

$$\sup_{1/4 \le t < 1/2} \int_{\mathrm{Re}\, z = t} |z|^2 \left| \hat{h}_j(z) \right|^2 dz \le C \left\| h_j, H^1(\Gamma_j) \right\|^2. \qquad (4.82)$$

The verification that $u \in E^+$ is therefore the same as in the Neumann case. Note that the condition $h_1(0) = h_2(0)$ has already been used to ensure the existence of a solution $u \in H^1$ of (4.37).   ■

Proof of Proposition 4.3.

- Firstly we consider the case $z = it$, $t \in \mathbb{R}$, $|t|$ large. We check the coercivity of the form q_z on H_z^1. (4.62) gives

$$\begin{cases} q_z(u, \bar{u}) = Q_t(u, \bar{u}) + O\left(\frac{\| u, H_{it}^1 \|^2}{t} \right) \\ Q_t(u, \bar{u}) = 2\mu \int \left(|\partial_\varphi u_\varphi|^2 + t^2 |u_r|^2 \right) d\varphi + \lambda \int |\partial_\varphi u_\varphi + itu_r|^2 d\varphi \quad (4.83) \\ \qquad + \mu \int |\partial_\varphi u_r + itu_\varphi|^2 d\varphi. \end{cases}$$

Setting $t = \frac{1}{h}$, $f = -iu_\varphi$, $g = u_r$, the coercivity of the form q_z results from the following lemma, which corresponds, at this stage, to the Korn inequality.   ■

Lemma 4.6. *Suppose that $I \subset \mathbb{R}$ is a bounded interval. There exist $c_0, h_0 > 0$ such that for any $h \in\,]0, h_0]$, $f, g \in H^1(I)$ we have*

$$\int_I |hf'|^2 + |g|^2 + |hg' - f|^2 \ge c_0 \int_I |f|^2. \qquad (4.84)$$

Proof. The proof of this lemma is left to the reader.

We deduce the following

Lemma 4.7. *For $z = it$, $t \in \mathbb{R}$, $|t|$ large, $T_z^{N,D}$ is an isomorphism, with inverse uniformly bounded with respect to t.*

Proof. We limit ourselves to the Neumann case, the Dirichlet case being quite similar. Because we know how to lift continuously the traces, and since q_{it} is coercive on H_{it}^1, it is sufficient to prove that the variational solution $u \in H_{it}^1$, given by the Lax-Milgram Lemma, of the problem

$$\begin{cases} A_{it}(u) = w \in L_{it}^2 \\ B_{it}(u) = 0 \ \text{ at } \varphi = 0, \ \varphi = \gamma \end{cases} \tag{4.85}$$

satisfies

$$\|u, H_{it}^2\| \le C_0 \|w, L^2\|, \tag{4.86}$$

where C_0 is independent of t. Since we have $q_{it}(u, v) = \int wv$ for $v \in H_{it}^1$ and since

$$\left| \int wv \right| \le \frac{C_0}{t} \|w; L^2\| \, \|v; H_{it}^1\|, \tag{4.87}$$

we have

$$\|u; H_{it}^1\| \le \frac{C_0}{t} \|w; L^2\|. \tag{4.88}$$

Therefore (4.86) results from (4.55) and from the equation $A_{it}(u) = w$. ∎

• We consider now the case $z = it + \sigma \in U$. For $|t|$ large, we have (see (4.55),(4.57))

$$\left\| T_z^{N,D} - T_{it}^{N,D} \right\| = O\left(\frac{1}{|t|} \right). \tag{4.89}$$

Therefore $T_z^{N,D}$ is an isomorphism for $z \in U$, $|z|$ large. In addition, since $T_{z_1} - T_{z_2}$ is compact, T_z is a Fredholm operator of index 0 for $z \in U$. The proof of the proposition is now reduced to the

Lemma 4.8. *For small values of $\alpha_0 > 0$ and $z \in U\backslash\{0\}$, the operator $T_z^{N,D}$ is injective.*

Proof. The solutions of the equation $A_z \begin{bmatrix} u_r \\ u_\varphi \end{bmatrix} = 0$ span a vector space of dimension 4. They are of the form (we note $T = T(z) = \frac{(a-b)z+(a+b)}{(a-b)z-(a+b)}$)

$$\begin{pmatrix} u_r \\ u_\varphi \end{pmatrix} = e^{i(z+1)\varphi} \begin{pmatrix} A \\ iA \end{pmatrix} + e^{-i(z+1)\varphi} \begin{pmatrix} B \\ -iB \end{pmatrix}$$
$$- e^{i(z-1)\varphi} \begin{pmatrix} C \\ iTC \end{pmatrix} + e^{-i(z-1)\varphi} \begin{pmatrix} -D \\ iTD \end{pmatrix} \tag{4.90}$$

Let us introduce the 2×2 matrices

$$\begin{cases} U^+ = \begin{pmatrix} e^{i\gamma(z+1)} & 0 \\ 0 & e^{-i\gamma(z+1)} \end{pmatrix} & U^- = \begin{pmatrix} e^{i\gamma(z-1)} & 0 \\ 0 & e^{-i\gamma(z-1)} \end{pmatrix} \\[2mm] N_1 = \begin{pmatrix} 1 & 1 \\ i & -i \end{pmatrix} & -N_2 = \begin{pmatrix} -1 & -1 \\ -iT & iT \end{pmatrix} \\[2mm] N_3 = 2ibz \begin{pmatrix} 1 & -1 \\ i & i \end{pmatrix} & -N_4 = -ib(1+t) \begin{pmatrix} z-1 & -(z-1) \\ i(z+1) & i(z+1) \end{pmatrix} \end{cases}$$

$$\tag{4.91}$$

A simple computation shows, using (4.57) and (4.90), that the solution u of the equation $A_z(u) = 0$ satisfies the Dirichlet (resp. Neumann) condition if and only if $M_D X = 0$ (resp. $M_N X = 0$), where $X = (A, B, C, D)$ and where the matrices $M_{D,N}$ are

$$M_D = \begin{pmatrix} N_1 & -N_2 \\ N_1 U^+ & -N_2 U^- \end{pmatrix}, \quad M_N = \begin{pmatrix} N_3 & -N_4 \\ N_3 U^+ & -N_4 U^- \end{pmatrix} \tag{4.92}$$

By setting

$$\mathcal{R}_D = N_1^{-1} N_2 = \frac{1}{2} \begin{pmatrix} 1+T & 1-T \\ 1-T & 1+T \end{pmatrix} \tag{4.93}$$

$$\mathcal{R}_N = N_3^{-1} N_4 = \frac{1+T}{2z} \begin{pmatrix} z & 1 \\ 1 & z \end{pmatrix} \tag{4.94}$$

the problem reduces to locate the roots of the equation

$$\det (U^+ \mathcal{R} - \mathcal{R} U^-) = 0. \tag{4.95}$$

By writing

$$\mathcal{R} = \begin{pmatrix} \alpha & \beta \\ \beta & \alpha \end{pmatrix} \tag{4.96}$$

we get

$$U^+ \mathcal{R} - \mathcal{R} U^- = 2i \begin{pmatrix} \alpha \sin \gamma \, e^{i\gamma z} & \beta \sin(\gamma z) e^{i\gamma} \\ -\beta \, e^{-i\gamma} \sin(\gamma z) & -\alpha \sin\gamma \, e^{-i\gamma z} \end{pmatrix}. \tag{4.97}$$

Therefore, (4.95) is equivalent to

$$(\sin(\gamma z))^2 = z^2 q^2 \sin^2 \gamma \tag{4.98}$$

with $q = \left| \frac{a-b}{a+b} \right| < 1$ in the Dirichlet case, and $q = 1$ in the Neumann case. Since

$$X_0 = \inf\{|\,\mathrm{Re}(z)|/z \text{ is a nonzero solution of (4.94)}\} \tag{4.99}$$

is the first real nonnegative root of (4.98), Lemma (4.8) results from $X_0 > \frac{1}{2}$.

■

4.2 Proof of Theorem 1. Existence and Uniqueness of the Spectral Function

Firstly, we prove the existence of the spectral function Σ. If $y = (y_1, y_2)$ is defined by Lemma 3.9, then $\Sigma = y + X$ is solution of $(3.31)_{\theta=0}$ if X is solution of the system

$$\begin{cases} DM.X_1 + TM.X_2 = u_1 \\ TM.X_1 + DM.X_2 = u_2. \end{cases} \qquad (4.100)$$

We deduce from Proposition 4.2 that (4.100) does have a (unique) solution in $\mathcal{H}$. By construction, $y \in \hat{\mathcal{A}}^6$, hence $\Sigma = y + X \in \hat{\mathcal{A}}^6$. The unicity of Σ in the class $\hat{\mathcal{A}}$ will result if

$$\begin{cases} DM.\Sigma_1 + TM.\Sigma_2 = 0 \\ TM.\Sigma_1 + DM.\Sigma_2 = 0 \end{cases} \qquad (4.101)$$

ensures that $\Sigma_1, \Sigma_2 = 0$. In view of Theorem 4.1, it is sufficient to check that a solution $\Sigma \in \hat{\mathcal{A}}^6$ of (4.2) can be extended analytically to the domain $\mathrm{Im}(\xi e^{-i\alpha}) < 0$ for $\alpha > 0$ small and is such that $\Sigma(\xi e^{i\alpha}) \in (H^+)^6$. Defining $g_j = TM.\Sigma_j$ we rewrite (4.101) as

$$DM.\Sigma_1 = -g_2; \quad DM.\Sigma_2 = -g_1. \qquad (4.102)$$

The application $\zeta \mapsto \cos\varphi\,\zeta + \sin\varphi\,\zeta_*(\zeta)$ (which is $D_*(\xi, \zeta)^{-1}$) maps the contour Γ_0 onto the contour pictured on Fig. 4.4. Therefore, there exists $\varepsilon_0 > 0$ such that

(i) If $f \in \hat{\mathcal{A}}^3$, then $(TM.f)(\nu_0 + \varepsilon_0 + \xi e^{i\alpha}) \in H_+^3$ for $\alpha \in [0, \varepsilon_0]$.

(ii) If $f \in \hat{\mathcal{A}}^3$ is holomorphic in a neighborhood of $[a, +\infty[$ for some $a \geq 0$, then $TM.f$ is holomorphic in a neighborhood of $[a - \varepsilon_0, +\infty[$.

It is therefore sufficient to check the two following properties of the operator DM (where $\varepsilon_0 > 0$ is small).

(1) If $g(\nu_0 + \varepsilon_0 + \xi e^{i\alpha}) \in H_+^3$ and $f \in \hat{\mathcal{A}}$, $DM.f = g$, then $f(\nu_0 + \varepsilon_0 + \xi e^{i\alpha}) \in H_+^3$.

(2) If g is holomorphic near $a \geq 0$ and if $f \in \hat{\mathcal{A}}$ is such that $DM.f = g$, then f is holomorphic near a.

To prove (1), we deform the contour Γ_0 onto the contour Γ_3, taking into account the residue at ξ

$$DM.f(\xi) = \int_{\Gamma_3} \frac{1}{2i\pi} \frac{1}{\xi - \zeta} \widetilde{dm}(\zeta).f(\zeta)d\zeta + \widetilde{dm}(\xi)f(\xi) = h(\xi) + \widetilde{dm}(\xi)f(\xi) \qquad (4.103)$$

where Γ_3 is the contour pictured on Fig. 4.5 and where $\widetilde{dm}$ is the holomorphic extension to Γ_3 of dm. We have that $h(\nu_0 + \varepsilon_0 + \xi e^{i\alpha}) \in H_+^3$ for $|\alpha|$ small and $\widetilde{dm}(\xi)$ is nonsingular with bounded inverse for $\xi = \nu_0 + \varepsilon_0 + \rho e^{i\alpha}$, $\rho \geq 0$, $0 \leq \alpha < \varepsilon_0$, for ξ lying off a neighborhood of ν_S. The property (1) follows, since $f \in \hat{A}$ ensures the holomorphy of f near ν_s.

In order to check (2), we may suppose $a \neq \nu_L, \nu_T, \nu_0, \nu_S$ because $f \in \hat{A}$. Hence the matrix $dm(\xi)$ is holomorphic and nonsingular near a and we have for $\xi \sim a$

$$DM.f(\xi) = \int_{\Gamma_a} \frac{1}{2i\pi} \frac{1}{\xi - \zeta} dm(\zeta).f(\zeta)d\zeta - dm(\xi)f(\xi) = h_a(\xi) - dm(\xi)f(\xi)$$

$$(4.104)$$

where Γ_a is the contour pictured on Fig. 4.6

Since h_a is holomorphic near a, we get the result. ∎

4.3 Proof of Theorem 2. Structure of the Spectral Function

The first part of Theorem 2 (structure of $\Sigma(\xi)$) results from Theorem 4.1. We have only to check the result about the boundary values $\Sigma^j(\xi_0 - i0^+)$, $\xi \in \mathbb{R}$.

(i) If $\xi_0 > -\nu_L$, the result is implied by the decomposition $\Sigma^j = y^j + X^j$, $X^j \in \mathcal{H}$, y^j meromorphic on $\mathbb{U}$ with poles in the set $\mathcal{P}^j$.

(ii) If $\xi_0 < -\nu_0$, then $(\xi_0 - i\varepsilon) \in \bigcap_* \Omega_*^-$ and $\operatorname{Im} T_*^{-1}(\xi_0 - i\varepsilon) < -c_0$ for $\varepsilon > 0$ small and $c_0 > 0$ independent of ε. In addition, the functions $dm(\xi).g_j$ are holomorphic near ξ_0. The result follows now from the functional equation (cf. Lemma 3.11) because the only root of $\delta(\xi - i0^+)$ for $\xi_0 < -\nu_0$ is $-\nu_S$, (cf 3.65).

(iii) We suppose $\xi_0 \in [-\nu_0, -\nu_L]$. We use again the functional equation (3.135), but with the following modification of the contour Γ_2. Recall that $TM(\xi, \zeta) = \frac{1}{2i\pi} \sum_* D_*(\xi, \zeta) tm_*(\zeta)$. For $* \in \{L, T, 0\}$, we choose three contours Γ_*, matching together at infinity and having the behaviour pictured on Fig. 4.7 . Defining $\mathbb{U}^*$ as the domain on the left of the contour $T_*(\Gamma_*)$ and the functions $\tilde{g}_1, \tilde{g}_2$ as

$$\begin{cases} \tilde{g}_1(\xi) = dm(\xi)^{-1} \left\{ \dfrac{W_1}{\xi - Z_1} - \displaystyle\int_{\Gamma_2} DM(\xi, \zeta) . \Sigma_1(\zeta) d\zeta \right. \\[2ex] \qquad\quad \left. - \displaystyle\sum_* \int_{\Gamma_*} \dfrac{1}{2i\pi} D_*(\xi, \zeta) . tm(\zeta) \Sigma_2(\zeta) d\zeta \right\} \\[3ex] \tilde{g}_2(\xi) = dm(\xi)^{-1} \left\{ \dfrac{W_2}{\xi - Z_2} - \displaystyle\int_{\Gamma_2} DM(\xi, \zeta) . \Sigma_2(\zeta) d\zeta \right. \\[2ex] \qquad\quad \left. - \displaystyle\sum_* \int_{\Gamma_*} \dfrac{1}{2i\pi} D_*(\xi, \zeta) . tm(\zeta) \Sigma_1(\zeta) d\zeta \right\} \end{cases} \qquad (4.105)$$

then, we have for $\xi \in (\text{Im } \xi < 0) \cap \mathbb{C} \backslash (\bigcup_* \partial\Omega_*)$ the modified functional equation

$$\begin{cases} \Sigma_1(\xi) = \tilde{g}_1(\xi) + \displaystyle\sum_* M_*(\xi) . \Sigma_2(T_*^{-1}(\xi)) \mathbf{1}(\xi \in \mathbb{U}_*) \\[2ex] \Sigma_2(\xi) = \tilde{g}_2(\xi) + \displaystyle\sum_* M_*(\xi) . \Sigma_1(T_*^{-1}(\xi)) \mathbf{1}(\xi \in \mathbb{U}_*) \end{cases} \qquad (4.106)$$

By a judicious choice of the contour Γ_*, we may assume that $\xi_0 \notin \{T_*(\zeta), \zeta \in \Gamma_*\}$. Therefore $g_j = dm(\xi)^{-1} f_j$ where the function f_j is holomorphic near ξ_0.

Finally, the matrix $dm(\xi)^{-1}$ behaves near ξ_0 according to (2.42). In addition, we may suppose that the structure of $\Sigma_j(\xi)$ is known near the points $T_*^{-1}(\xi_0)$. Therefore, the second part of theorem 2 is a consequence of (4.106) and of the formulas defining the matrices $M_*(\xi)$. ∎

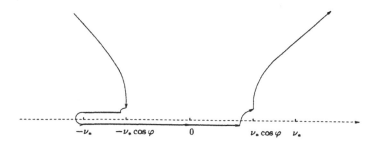

Fig. 4.4.1. The image of the contour Γ_0 by the map $\zeta \mapsto \cos \varphi \, \zeta + \sin \varphi \, \zeta_*(\zeta)$, when $\varphi < \pi/2$.

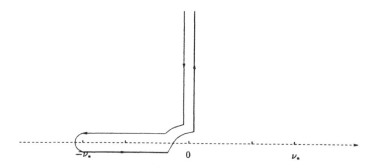

Fig. 4.4.2. The image of the contour Γ_0 by the map $\zeta \mapsto \cos \varphi \, \zeta + \sin \varphi \, \zeta_*(\zeta)$, when $\varphi = \pi/2$.

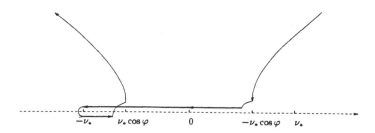

Fig. 4.4.3. The image of the contour Γ_0 by the map $\zeta \mapsto \cos \varphi \, \zeta + \sin \varphi \, \zeta_*(\zeta)$, when $\varphi > \pi/2$.

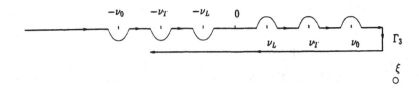

Fig. 4.5. The contour Γ_3

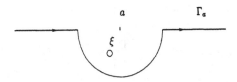

Fig. 4.6. The contour Γ_a

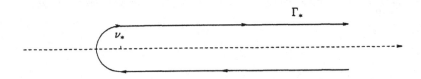

Fig. 4.7. The contour Γ_*

5. Numerical Algorithm

5.1 Introduction

The aim of this section is to derive from the mathematical analysis of Sect.2, Sect.3, Sect.4 an efficient numerical algorithm for computing accurately the spectral function Σ. If $\theta_{\text{obs}} \in\ ] - (\pi - \varphi), \pi[$ is a given angle of observation (Fig. 2.1), the amplitude $A(\theta_{\text{obs}})$ of the diffracted wave in the fluid is asymptotically given by (§2.7)

$$A(\theta_{\text{obs}}) \sim |\hat{\gamma}_1(\nu_0 \cos \theta_{\text{obs}} - i0^+) + \hat{\gamma}_2(\nu_0 \cos(\varphi - \theta_{\text{obs}}) - i0^+)| \qquad (5.1)$$

where $\hat{\gamma}_j$ is the third component of the spectral function Σ_j. Therefore, we have to compute an uniformly accurate approximation of the spectral function on the segment $] - \nu_0, \nu_0[-i0^+$.

The algorithm presented hereafter, is based on the decomposition of the spectral function in the form $\Sigma(\xi) = y(\xi) + X(\xi)$. Three major ingredients are used:

- the explicit computation of the part $y(\xi)$.
- the computation of $X(\xi)$ by a Galerkin-collocation method.
- the evaluation of $\Sigma(\xi)$ with the help of the functional equation (Sect.3.5).

Two cases of incident waves have been considered. The first is the one of an incident plane wave in the fluid, making an angle $\theta_{in} \in\ ]0, \pi - \frac{\varphi}{2}[$ with the face 1 of the wedge. This case is the one considered until now in the previous theoretical study. The second case is the one of an incident Scholte-Stoneley wave (§3.2) propagating along the face 1 of the wedge, from the right towards the vertex. In this last case, the right-hand-side in the integral system (3.37) has to be modified. Note finally that these two cases are of physical interest and give rise to numerous experiments in the acoustic laboratories.

5.2 The Case of an Incident Plane Wave in the Fluid

In this case, the spectral function $\Sigma(\xi) = (\Sigma_1(\xi), \Sigma_2(\xi))$ is solution of the system $(3.31)_{\theta=0}$. Using the symmetry of the problem, we may assume $\theta_{in} \in]0, \pi - \frac{\varphi}{2}[$. The algorithm of computation of the function $y_1(\xi), y_2(\xi)$ given in §3.4 is detailed now. We distinguish two cases.

1st case : the two faces are illuminated

This case corresponds to $\theta_{in} \in]\pi - \varphi, \pi - \frac{\varphi}{2}[$. If $\theta_0^1, \theta_0^2 \in \mathcal{D}$ (see Sect.2.4 for the definition of $\mathcal{D}$) are the two complex angles such that $Z_1 = \nu_0 \cos \theta_0^1$, $Z_2 = \nu_0 \cos \theta_0^2$, we have $\theta_0^1 = \theta_{in}$, $\theta_0^2 = 2\pi - (\theta_{in} + \varphi)$. The points $Z_1, Z_2 \notin \Omega_0^+$, therefore they have no iterates of type "0". Again there are two cases.

(i) $0 < \varphi \le \frac{\pi}{2}$.

We have $\Omega_L^+ \subset \Omega_T^+ \subset \Omega_0^+$, therefore $Z_{1,2} \notin \Omega_{L,T}^+$ and $Z_{1,2}$ have also no iterate of type L or T.

(ii) $\frac{\pi}{2} < \varphi < \pi$.

We have $\Omega_0^+ \subset \Omega_T^+ \subset \Omega_L^+ \subset \{\text{Re}\, z \ge 0\}$. The poles $Z_{1,2}$ can have an iterate of type L or T. For example, if $Z_1 = \nu_L \cos \theta_L^1$, $0 < \text{Re}\, \theta_L^1 < \pi - \varphi$, $T_L(Z_1) = \nu_L \cos(\theta_L^1 + \varphi)$ with $\text{Re}(\theta_L^1 + \varphi) > \frac{\pi}{2}$, which implies $T_L(Z_1) \notin \Omega_{L,T,0}^+$. Therefore there are no longer iterates. The same property holds for $T_T(Z_1)$. Consequently there are at most 2 generations of poles $Z_\ell = Z_\ell^1 \cup Z_\ell^2, 0 \le \ell \le 1$ with

- $Z_0^1 = \{Z_1\}$, $Z_0^2 = \{Z_2\}$
- $Z_1^1 = \{T_L(Z_2), T_T(Z_2)\}$; $Z_1^2 = \{T_L(Z_1), T_T(Z_1)\}$

where the sets $Z_1^j, j = 1, 2$ are eventually empty. The residues corresponding to the poles $z \in Z_\ell^j$ are the vectors $w(z) \in \mathbb{C}^3$ given by (cf. (3.115-3.118))

- $\ell = 0 : w(Z_1) = dm^{-1}(Z_1) \cdot W_1$; $w(Z_2) = dm^{-1}(Z_2) \cdot W_2$
- $\ell = 1 : w(z) = -\tilde{t}_*(z') \cdot w(z')$ where z' is given by $T_*(z') = z$.

2cd case : one face is illuminated

This case corresponds to $\theta_{in} \in]0, \pi - \varphi[$. Because of $Z_1 = \nu_0 \cos \theta_{in}$, $Z_2 = \nu_0 \cos(\theta_{in} + \varphi)$, we see that $Z_1 \in \Omega_0^+$ and that $Z_2 = T_0(Z_1)$. Therefore, the recurrence relations for building the two generations of poles can be restricted to the construction of the iterates of the generation Z_1. In addition, due to (3.119), they are no iterates of type L or T of a pole of type 0, (because the residue is 0). Consequently the recurrence relation is

- Generation 0: $Z_0 = \{Z_1\}$, with residue w

- Generation 1: $Z_1 = \{T_L(Z_1 \cap \Omega_L^+), T_T(Z_1 \cap \Omega_T^+), T_0(Z_1) = Z_2\}$

- for $\ell \geq 1$, we note $Z_{\ell,0} \subset Z_\ell$ the subset of the iterates which are themselves of type 0. The generation $Z_{\ell+1}$ is given by

$$Z_{\ell+1} = T_L\big((Z_\ell \cap \Omega_L^+) \setminus Z_{\ell,0}\big) \cup T_T\big((Z_\ell \cap \Omega_T^+) \setminus Z_{\ell,0}\big) \cup T_0(Z_\ell \cap \Omega_0^+) \quad (5.2)$$

and the residues are

- $z \in Z_0 : w(z = Z_1) = dm^{-1}(Z_1) \cdot W_1$

- $z \in Z_1$: for $z = T_{L,T}(Z_1 \cap \Omega_{L,T}^+)$, $w(z) = -\tilde{t}_{L,T}(Z_1)w(Z_1)$ and $w(z = Z_2) = dm^{-1}(Z_2) \cdot W_2 - \tilde{t}_0(Z_1)w(Z_1)$

- $z \in Z_\ell, \ell \geq 2$: for $z = T_*(z')$, $z' \in Z_\ell$, $w(z) = -\tilde{t}_*(z') \cdot w(z')$.

The functions $y_1(\xi)$, $y_2(\xi)$, given by Lemma 3.9, are therefore

$$y_1(\xi) = \sum_{\ell \text{ even}} \sum_{z \in Z_\ell} \frac{w(z)}{\xi - z}; \quad y_2(\xi) = \sum_{\ell \text{ odd}} \sum_{z \in Z_\ell} \frac{w(z)}{\xi - z}. \quad (5.3)$$

5.3. The Case of an Incident Scholte-Stoneley Wave

We consider in this section the case of an incident Scholte-Stoneley wave along the face 1 of the wedge, which propagates from the right to the left, (Fig. 5.1). The Scholte-Stoneley wave is an interface-wave type, solution of the system $(S_{\theta=0})$ given by (3.39). It reads

$$v_S(x,y) = e^{ix\nu_S}\overline{v}_S(y), \ y > 0; \quad h_S(x,y) = e^{ix\nu_S}\overline{h}_S(y), \ y < 0 \quad (5.4)$$

where $\overline{v}_S(y)$, $\overline{h}_S(y)$ are the rapidly decreasing functions

$$
\begin{cases}
\overline{v}_S(y) = \dfrac{i}{2}\Big\{ e^{iy\zeta_L(\nu_S)}\Big(\hat{\beta}_S + \dfrac{\nu_S}{\zeta_L(\nu_S)}\hat{\alpha}_S\Big)\begin{bmatrix} \nu_S \\ \zeta_L(\nu_S) \end{bmatrix} \\
\qquad\quad + e^{iy\zeta_T(\nu_S)}\Big(-\hat{\alpha}_S + \dfrac{\nu_S}{\zeta_T(\nu_S)}\hat{\beta}_S\Big)\begin{bmatrix} -\zeta_T(\nu_S) \\ \nu_S \end{bmatrix}\Big\}, \ y > 0 \\[2mm]
\overline{h}_S(y) = \dfrac{i}{2}e^{-iy\zeta_0(\nu_S)}\dfrac{\hat{\gamma}_S}{\zeta_0(\nu_S)}, \ y < 0
\end{cases}
$$

$$\qquad\qquad\qquad\qquad\qquad\qquad\qquad\qquad\qquad\qquad (5.5)$$

$$\zeta_{L,T,0}(\nu_S) = i\sqrt{\nu_S^2 - \nu_{L,T,0}^2} \quad (5.6)$$

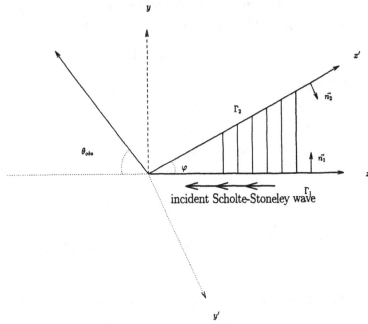

Fig. 5.1. Incident Scholte-Stoneley wave

and where $(\hat{\alpha}_S, \hat{\beta}_S, \hat{\gamma}_S)^T$ is any non zero vector solution of

$$dm(\nu_S) \begin{bmatrix} \hat{\alpha}_S \\ \hat{\beta}_S \\ \hat{\nu}_S \end{bmatrix} = 0. \tag{5.7}$$

We look now for a solution $\tilde{v}(x,y)$, $\tilde{h}(x,y)$ of the system $(3.39)_{\theta=0}$, where the functions $(\tilde{v}, \tilde{h})$ contain an incident Scholte-Stoneley wave along the face 1. The formulation of the problem cannot be exactly the same as in the case of an incident wave in the fluid, because the interface Scholte-Stoneley wave (5.4-5.5) is not a solution of the wave equation in the fluid. Therefore we leave *in the unknown* the incident Scholte-Stoneley wave in the following way. Going back to the system (2.3-2.6), the system to solve is

$$
\begin{cases}
(E+1)\tilde{v} = 0 & \text{in } \Omega_s \\
(\Delta + \nu_0^2)\tilde{h} = 0 & \text{in } \Omega_s \\
(\lambda \operatorname{div} \tilde{v} + 2\mu\varepsilon(\tilde{v}))\mathbf{n} - i\rho\tilde{h}\mathbf{n} = 0 & \text{on } \Gamma \\
i\tilde{v} \cdot \mathbf{n} - \operatorname{grad} \tilde{h} \cdot \mathbf{n} = 0 & \text{on } \Gamma.
\end{cases}
\tag{5.8}
$$

Suppose that $w_s = (\hat{\alpha}_S, \hat{\beta}_S, \hat{\gamma}_S)^T \in \mathbb{C}^3$ is a vector such that $dm(\nu_S) \cdot w_S = 0$, and $v_S(x,y)$, $h_S(x,y)$ is a Scholte-Stoneley interface wave (5.4), an *outgoing solution* of (5.8) is defined by

Definition 5.1 (Outgoing solution for a Scholte-Stoneley incident wave). *An outgoing solution of the system* (5.8), *containing an incident Scholte-Stoneley wave along the face 1 of the wedge is a solution* (v,h) *of the form*

$$
v = v_1|_{\Omega_s} + v_2|_{\Omega_s}; \quad h = h_1|_{\Omega_f} + h_2|_{\Omega_f}
\tag{5.9}
$$

with

$$
v_j = \lim_{\varepsilon \to 0^+} -(E + e^{-2i\varepsilon})^{-1} \left[\begin{pmatrix} \alpha_j \\ \beta_j \end{pmatrix} \otimes \delta_j \right]
\tag{5.10}
$$

$$
h_j = \lim_{\varepsilon \to 0^+} -(\Delta + \nu_0^2 e^{-2i\varepsilon})^{-1} [\gamma_j \otimes \delta_j].
\tag{5.11}
$$

where the spectral functions $\Sigma_j = (\hat{\alpha}_j, \hat{\beta}_j, \hat{\gamma}_j)^T$ *are such that*

$$
\Sigma_1 \in \operatorname{Span}\left(\frac{w_S}{\xi - \nu_S} \right) + \hat{\mathcal{A}}^3, \qquad \Sigma_2 \in \hat{\mathcal{A}}^3.
\tag{5.12}
$$

The constant C_0 appearing in Definition 2.1 of the class $\mathcal{A}$ is taken such that $C_0 > \nu_S$. Lemma 2.2 still holds, but in the final representation formulas (2.57-2.58), the singularity ν_S should be contoured from below by Γ_0 as shown on Fig. 5.2. Since the null space of the matrix $dm(\nu_S)$ is one-dimensional, the vector w_S is unique up to a constant. We choose $w_S = (w_1, w_2, w_3)$ as

$$
\begin{cases}
w_1 = w_2 \, \dfrac{\nu_S}{i\sqrt{\nu_S^2 - \nu_T^2}} \left[1 + 2\mu \left(\sqrt{\nu_S^2 - 1} \, \sqrt{\nu_S^2 - \nu_T^2} - \nu_S^2 \right) \right] \\[3mm]
w_2 = w_3 \, \dfrac{i\sqrt{\nu_S^2 - \nu_T^2}}{\sqrt{\nu_S^2 - 1} \, \sqrt{\nu_S^2 - \nu_T^2} - \nu_S^2} \\[3mm]
w_3 = \sqrt{\nu_S^2 - \nu_0^2}.
\end{cases}
\tag{5.13}
$$

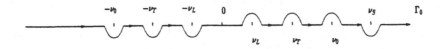

Fig. 5.2. The contour Γ_0 in the case of an incident Scholte-Stoneley wave

As in the second case of Sect.5.2, the functions $y_1(\xi)$, $y_2(\xi)$ are determined by only one sequence of poles, ordered in generations Z_ℓ, $\ell \geq 0$. For $\ell \geq 1$, we note still $Z_{\ell,0}$ the subset of the poles which are themselves iterates of type "0". We have (see the second case in Sect.5.2)

- $Z_0 = \{\nu_S\}$, $w(\nu_S) = w_S$
- $Z_{\ell+1} = T_L\big((Z_\ell \cap \Omega_L^+) \setminus Z_{\ell,0}\big) \cup T_T\big((Z_\ell \cap \Omega_T^+) \setminus Z_{\ell,0}\big) \cup T_0(Z_\ell \cap \Omega_0^+)$.

For $z \in Z_{\ell+1}$, $z = T_*(z')$, $z' \in Z_\ell$, the residue of z is given by

$$w(z) = -\tilde{t}_*(z') \cdot w(z'). \tag{5.14}$$

Defining now the poles of the faces 1 and 2 by $Z_1 = \underset{\substack{\ell \geq 0 \\ \ell \text{even}}}{\cup} Z_\ell$, $Z_2 = \underset{\substack{\ell \geq 0 \\ \ell \text{odd}}}{\cup} Z_\ell$, the y−part of $\Sigma_j(\xi)$ are

$$y_1(\xi) = \sum_{z \in Z_1} \frac{w(z)}{\xi - z}; \qquad y_2(\xi) = \sum_{z \in Z_2} \frac{w(z)}{\xi - z}. \tag{5.15}$$

Due to the fact that $\nu_S > \nu_*$, $* \in \{L, T, 0\}$, the three values $T_*(\nu_S)$ are in the half-plane $\text{Im}\, z > 0$. Thus they can not generate ν_L, ν_T, ν_0, ν_S as iterates.

Our existence and uniqueness result for the problem in the Scholte-Stoneley case is

Theorem 1$_S$. *The system* (5.8) *does have an unique outgoing solution* (v, h). *The corresponding spectral function is solution of*

$$\begin{cases} DM \cdot \Sigma_1 + TM \cdot \Sigma_2 = 0 \\ TM \cdot \Sigma_1 + DM \cdot \Sigma_2 = 0. \end{cases} \tag{5.16}$$

In theorem 2(Sect.2.5), we had to take care of the poles located on the real axis. In particular, the iterates of the branch points $-\nu_L$, $-\nu_T$ and the point $-\nu_0$ could coincide with poles. Here, we check easily that there is no pole on the real axis. Hence, the set C defined as the union of the iterates of $-\nu_L$ and $-\nu_T$, does not intersect the set of the poles $Z = \bigcup_{\ell \geq 0} Z_\ell$.

Theorem 2 on the structure of the spectral function is replaced here by

Theorem 2$_S$.

(i) For $j = 1, 2$, the function Σ_j is meromorphic on $\mathbb{U}$. We have the decomposition

$$\Sigma_j(\xi) = y_j(\xi) + X_j(\xi), \qquad j = 1, 2 \tag{5.17}$$

where $X_j \in \mathcal{H}$ and y_j is given by (5.15).

(ii) The boundary values $\Sigma_j(\xi - i0)$, $\xi \in \mathbb{R}$ are analytical for $\xi \notin C \cup \{-\nu_S\}$. Moreover $\Sigma_j(\xi - i0)$ has only simple poles in $\{-\nu_S, \nu_S\}$. In a neighborhood of ξ_0, the following equality holds

$$\Sigma_j(\xi - i0) = a(\xi) + (\xi - \xi_0)^{1/2} b(\xi) \tag{5.18}$$

where $a(\xi)$, $b(\xi)$ are holomorphic near ξ_0.

The functions (v_1, h_1) are given by (2.57-2.58) where $\varepsilon = 0$. Applying the Cauchy formula to the term $\frac{w}{\xi - \nu_S}$ in (2.57-2.58) gives that the Scholte-Stoneley wave (5.5) is present in the solution with the coefficients $(\hat{\alpha}_S, \hat{\beta}_S, \hat{\gamma}_S) = i(w_1, w_2, w_3)$.

5.4. Approximation of the Regular Part of the Spectral Function

In this paragraph, we describe the numerical method for the approximation of the system

$$\begin{cases} DM \cdot \Sigma_1(\xi) + TM \cdot \Sigma_2(\xi) = S_1(\xi) \\ TM \cdot \Sigma_1(\xi) + DM \cdot \Sigma_2(\xi) = S_2(\xi) \end{cases} \tag{5.19}$$

The right-hand side $S_j(\xi) = \frac{W_j}{\xi - Z_j}$, in the case of an incident wave in the fluid, and $S_j(\xi) = 0$ in the case of an incident Scholte-Stoneley wave along the face 1 of the wedge. Subtracting from each line of (5.19) respectively the functions

$$DM \cdot y_1(\xi) + TM \cdot y_2(\xi) \quad , \quad TM \cdot y_1(\xi) + DM \cdot y_2(\xi), \qquad (5.20)$$

we obtain the following system in $X_1(\xi)$, $X_2(\xi)$

$$\begin{cases} DM \cdot X_1(\xi) + TM \cdot X_2(\xi) = u_1(\xi) \\ TM \cdot X_1(\xi) + DM \cdot X_2(\xi) = u_2(\xi) \end{cases} \qquad (5.21)$$

where the functions $u_1, u_2 \in \mathcal{H}^3$ are given by

$$\begin{cases} u_1(\xi) = S_1(\xi) - \left[\sum_{z \in Z_1} D_p(z, w(z))(\xi) + \sum_{z \in Z_2} T_p(z, w(z))(\xi) \right] \\ u_2(\xi) = S_2(\xi) - \left[\sum_{z \in Z_1} T_p(z, w(z))(\xi) + \sum_{z \in Z_2} D_p(z, w(z))(\xi) \right]. \end{cases} \qquad (5.22)$$

We know from Proposition 4.2 that (5.21) does have an unique solution $(X_1, X_2) \in \mathcal{H}^3 \oplus \mathcal{H}^3$. Due to the absence of explicit formula for these functions, we approximate now (5.21) by a Galerkin-collocation method.

We choose $V = \text{Span}(\varphi_k)_{1 \le k \le N}$ a subspace of $\mathcal{H}$ of dimension $N \ge 1$, as Galerkin space for the approximations of $X_1(\xi)$, $X_2(\xi)$. Let $(b_\ell)_{1 \le \ell \le N}$, $\text{Im } b_\ell < 0$ be collocation points. We look for vectors $(X_1^k)_{1 \le k \le N}, (X_2^k)_{1 \le k \le N} \in \mathbb{C}^3$ such that the functions

$$\overline{X}_1(\xi) = \sum_{k=1}^N X_1^k \varphi_k(\xi), \qquad \overline{X}_2(\xi) = \sum_{k=1}^N X_2^k \varphi_k(\xi) \qquad (5.23)$$

are solutions of the collocation system

$$\begin{cases} DM \cdot \overline{X}_1(b_\ell) + TM \cdot \overline{X}_2(b_\ell) = u_1(b_\ell) \\ TM \cdot \overline{X}_1(b_\ell) + DM \cdot \overline{X}_2(b_\ell) = u_2(b_\ell). \end{cases} \qquad (5.24)$$

We note $[X_1], [X_2], [U_1], [U_2] \in \mathbb{C}^{3N}$ the vectors with components

$$[X_\alpha] = \begin{bmatrix} \overline{X}_\alpha^1 \\ \vdots \\ \overline{X}_\alpha^N \end{bmatrix}, \quad \overline{X}_\alpha^k \in \mathbb{C}^3; \quad [U_\alpha] = \begin{bmatrix} u_\alpha(b_1) \\ \vdots \\ u_\alpha(b_N) \end{bmatrix}, \quad \alpha = 1, 2. \qquad (5.25)$$

We note moreover $[D], [T] \in M_{3N}(\mathbb{C})$ the block matrices

$$[D] = \begin{bmatrix} D_{11} & \cdots & D_{1N} \\ \vdots & & \vdots \\ D_{N1} & \cdots & D_{NN} \end{bmatrix}, \quad [T] = \begin{bmatrix} T_{11} & \cdots & T_{1N} \\ \vdots & & \vdots \\ T_{N1} & \cdots & T_{NN} \end{bmatrix} \quad (5.26)$$

where the matrices $D_{\ell k}$, $T_{\ell k}$ are the 3×3 matrices defined by

$$[D_{\ell k}]_{ij} = (DM_{ij} \cdot \varphi_k)(b_\ell), \quad [T_{\ell k}]_{ij} = (TM_{ij} \cdot \varphi_k)(b_\ell), \quad 1 \le i, j \le 3.$$

The discrete system (5.24) can be rewritten as

$$\begin{bmatrix} [D] & [T] \\ [T] & [D] \end{bmatrix} \begin{bmatrix} [X_1] \\ [X_2] \end{bmatrix} = \begin{bmatrix} [U_1] \\ [U_2] \end{bmatrix} \quad (5.27)$$

or, equivalently

$$\begin{cases} ([D] + [T])([X_1] + [X_2]) = [U_1] + [U_2] \\ ([D] - [T])([X_1] - [X_2]) = [U_1] - [U_2]. \end{cases} \quad (5.28)$$

A natural choice of the Galerkin basis φ_k, $1 \le k \le N$ is based on the following representation formula. For $f \in \mathcal{H}$, such that the jump $[f(X)] = (f(x + i0^+) - f(x + i0^-))\mathbb{1}(]-\infty, -1])(x) \in L^2(]-\infty, -1])$, we have, by the Cauchy formula

$$f(\xi) = \frac{1}{2i\pi} \int_1^{+\infty} \frac{-[f(-\zeta)]}{\zeta + \xi} d\zeta. \quad (5.29)$$

The approximation of this integral at the points $e_k \ge 1$, suggests to approximate $f(\xi)$ with the Galerkin basis φ_k, $1 \le k \le N$ given by

$$\varphi_k(\xi) = \frac{d_k}{\xi + e_k}, \quad d_k \in \mathbb{C}. \quad (5.30)$$

The functions $X_j(\xi) \in \mathcal{H}$ (Sect.3.3) and are analytic at each $\xi \in \mathbb{R}$ with the possible exception of $\xi = -\nu_S$ (cf. Theorem 2, Sect.2.5). (If it is present in the *outgoing solution*, the function $\frac{1}{\xi + \nu_S}$ is in the X−part of the spectral function). Therefore, we can apply the formula (5.29) and take as a Galerkin basis the functions $\varphi_k(\xi)$, with the adjonction of the function $\frac{1}{\xi + \nu_S}$. The collocation points are taken as $b_k = e_k + i0^-$. This choice ensures that the interpolation operator by this Galerkin-collocation method does have an unique solution. In fact, suppose that $\overline{X}(\xi) = \sum_{k=1}^{N} X^k \varphi_k(\xi)$ is such that $\overline{X}(b_\ell) = 0$,

$1 \le \ell \le N$. Then, the polynomial $P(\xi) = \sum_{k=1}^{N} X^k \prod_{k' \neq k} (\xi + e_{k'})$ is of degree $N - 1$ and admits the N points b_ℓ as zero. Hence $P(\xi)$ is zero, therefore $X^k = 0$, $1 \le k \le N$.

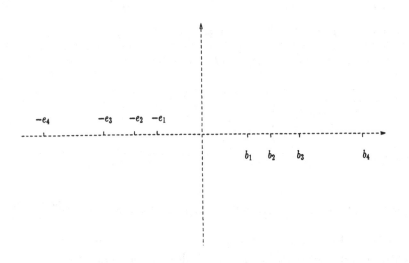

Fig. 5.3. The Galerkin points $-e_k$ and the collocation points b_k

The drawback of the foregoing choice of the Galerkin basis is that the approximations $\overline{X}_j(\xi)$ can not be uniformly accurate for $\text{Im}\,\xi < 0$. In particular, due to the non boundedness of $\varphi_k(\xi)$ near $\xi = -e_k$, $\overline{X}_j(\xi)$ can not be a good approximation of $X_j(\xi)$ on the segment $[-\nu_0, -\nu_L]$. Therefore, the evaluation of the spectral function $\Sigma_\alpha(\xi)$ by the "direct" formula

$$\Sigma_\alpha(\xi) = y_\alpha(\xi) + \overline{X}_\alpha(\xi), \qquad \alpha = 1, 2 \tag{5.31}$$

can no longer be good in a neighborhood of $-e_k$.

Another Galerkin basis which does not suffer from this hindrance is obtained by taking

$$\varphi_k(\xi) = \frac{1}{2i\pi} \int_1^{+\infty} \frac{\chi_k(\zeta)}{\zeta + \xi} d\zeta \tag{5.32}$$

with $\chi_k(\xi)$ is the *hat* function centered at e_k and of half width $\delta > 0$

$$\chi_k(\zeta) = \left(1 - \left|\frac{\zeta - e_k}{\delta}\right|\right) \mathbb{1}\left(\left|\frac{\zeta - e_k}{\delta}\right| \le 1\right). \tag{5.33}$$

We have, with $z = \frac{e_k + \xi}{\delta}$

$$\varphi_k(\xi) = (1+z)\ln\left(1+\frac{1}{z}\right) - (1-z)\ln\left(1-\frac{1}{z}\right). \tag{5.34}$$

This times, the function φ_k remains bounded in a neighborhood of $-e_k$. However, the numerical experiments have proved that this is not sufficient to make better the approximation of $X_\alpha(\xi)$ in the neighborhood of $]-\infty, -1]$. A solution to this problem is given in Sect.5.5.

Assembling the linear systems (5.28) proceeds in the following way. Recall that the integral system to solve is

$$\begin{cases} \int_{\Gamma_0} DM(\xi,\zeta) \cdot X_1(\zeta)d\zeta + \int_{\Gamma_0} TM(\xi,\zeta) \cdot X_2(\zeta)d\zeta = u_1(\xi) \\ \int_{\Gamma_0} TM(\xi,\zeta) \cdot X_1(\zeta)d\zeta + \int_{\Gamma_0} DM(\xi,\zeta) \cdot X_2(\zeta)d\zeta = u_2(\xi). \end{cases} \tag{5.35}$$

Since the functions $X_1, X_2 \in \mathcal{H}^3$, we may deform in (5.35) the contour Γ_0 onto the contour $\tilde{\Gamma}_0 = \{iy, y \in \mathbb{R}\}$. This is equivalent to make in each integral the change of variables $\zeta = iy$. By denoting $\xi = ix$ and

$$\begin{cases} \tilde{X}_\alpha(y) = X_\alpha(iy) \\ \tilde{u}_\alpha(x) = u_\alpha(ix) \\ \widetilde{DM}(x,y) = iDM(ix, iy) \\ \widetilde{TM}(x,y) = iTM(ix, iy) \end{cases} \tag{5.36}$$

the system (5.35) becomes

$$\begin{cases} \int_{\mathbb{R}} \widetilde{DM}(x,y) \cdot \tilde{X}_1(y)dy + \int_{\mathbb{R}} \widetilde{TM}(x,y) \cdot \tilde{X}_2(y)dy = \tilde{u}_1(x) \\ \int_{\mathbb{R}} \widetilde{TM}(x,y) \cdot \tilde{X}_1(y)dy + \int_{\mathbb{R}} \widetilde{DM}(x,y) \cdot \tilde{X}_2(y)dy = \tilde{u}_2(x). \end{cases} \tag{5.37}$$

The kernels $\widetilde{DM}(x,y), \widetilde{TM}(x,y)$ are now elliptic. In particular, they have no singular points. The Galerkin space for the approximation of $\tilde{X}_1, \tilde{X}_2$ is generated by $\tilde{\varphi}_k(x) = \varphi_k(ix)$, i.e.

$$\tilde{\varphi}_k(x) = \frac{c_k}{x - ie_k}. \tag{5.38}$$

The collocation points b_ℓ become $-ib_\ell$ in the new coordinates. The coefficients $[D_{\ell k}]_{ij}$, $[T_{\ell k}]_{ij}$, $u_{1,2}(b_\ell)$ occuring in the two linear systems (5.28) are

$$[D_{\ell k}]_{ij} = (\widetilde{DM}_{ij} \cdot \tilde{\varphi}_k)(-ib_\ell), \quad [T_{\ell k}]_{ij} = (\widetilde{TM}_{ij} \cdot \tilde{\varphi}_k)(-ib_\ell) \tag{5.39}$$

$$u_{1,2}(b_\ell) = \tilde{u}_{1,2}(-ib_\ell). \tag{5.40}$$

5.5. Computation of the Spectral Function

According to Sect.2.6, we wish now to compute the diagram of diffraction of the wedge, that is the function

$$\theta_{obs} \in]-(\pi - \varphi), \pi[\mapsto |\Sigma_1(\nu_0 \cos\theta_{obs} + i0^-) + \Sigma_2(\nu_0 \cos(\varphi - \theta_{obs}) + i0^-)|. \tag{5.41}$$

The *direct* evaluation of the spectral function is simply to approximate $\Sigma_{1,2}(\xi)$ by

$$\Sigma_{1,2}(\xi) \simeq y_{1,2}(\xi) + \overline{X}_{1,2}(\xi), \qquad \mathrm{Im}\,\xi < 0. \tag{5.42}$$

where the $y-$part $y_{1,2}(\xi)$, is given by (5.3), (incident wave in the fluid), or (5.15), (incident Scholte-Stoneley wave) and the approximation $\overline{X}_{1,2}(\xi)$ of $X_{1,2}(\xi)$, is known by the resolution of the two linear systems (5.28). This formula has been proved to be very accurate for ξ in a neighborhood of the collocation points b_ℓ, in practice for ξ such that, say, $\mathrm{Re}\,\xi \geq \nu_L = 1$. On the contrary, $\overline{X}_{1,2}(\xi)$ becomes inaccurate when ξ lies in the quadrant $\mathrm{Im}\,\xi < 0$, $\mathrm{Re}\,\xi < \nu_L$. The remedy is to use again the functional equation

$$\begin{cases} \Sigma_1(\xi) = g_1(\xi) + \displaystyle\sum_{*\in\{L,T,0\}} M_*(\xi)\Sigma_2(T_*^{-1}(\xi))\mathbb{1}(\xi \in \Omega_*^-) \\[2mm] \Sigma_2(\xi) = g_2(\xi) + \displaystyle\sum_{*\in\{L,T,0\}} M_*(\xi)\Sigma_1(T_*^{-1}(\xi))\mathbb{1}(\xi \in \Omega_*^-) \end{cases} \tag{5.43}$$

where g_1, g_2 are given by (3.134) and $M_*(\xi)$ are the transfer matrices $M_*(\xi)$ (3.132).

Because of the fact that $M_{L,T}(\zeta)M_0(T_{L,T}^{-1}(\zeta)) = 0$ (see (3.136-3.138)), the iterate of type 0 of a point ζ of type L or T gives a null contribution in the recursive equation (5.43). Therefore, the computation of these iterates can be skipped. The sequence of the iterates generated by the observation point ξ can be ordered in generations Z_ℓ, $\ell \geq 0$ (note the analogy with the sequence of the poles)

$$\ell = 0, \qquad Z_0 = \tilde{z}_0 = \{\xi\}$$
$$\ell \geq 0, \qquad Z_{\ell+1} = \tilde{z}_{\ell+1} \cup T_L^{-1}(Z_L \cap \Omega_L^-) \cup T_T^{-1}(Z_\ell \cap \Omega_T^-) \tag{5.44}$$
$$\text{with } \tilde{z}_{\ell+1} = T_0^{-1}(\tilde{z}_\ell \cap \Omega_0^-).$$

The functions $\xi = \nu_* \cos\theta \in \Omega_*^- \mapsto T_*^{-1}(\xi) = \nu_* \cos(\theta - \varphi)$ maps the points from the left to the right. As in Lemma 2.4, we can prove that the sequence Z_ℓ is finite, and that the terminal points are located in the region of admissibility of (5.43). Thus, (5.44) provides *a priori* a good method to propagate the accuracy of the direct formula from the quadrant $\text{Im}\,\xi < 0$, $\text{Re}\,\xi > \nu_L = 1$ to the whole half-space $\text{Im}(\xi) < 0$. The numerical experiments have confirmed this view.

5.6. Practical Issues

We summarize here the algorithm of computation of $\Sigma_{1,2}(\xi)$ and indicate some details of the effective implementation.

a) Computation of the $y-$part

The poles and the residues are computed by the algorithms described in Sect.5.2 for an incident wave in the fluid, and in Sect.5.3 for an incident Scholte-Stoneley wave.

b) Computation of the $X-$part

We compute first the coefficients given by (5.39-5.40) of the matrices $[D]$, $[T]$ and of the right-hand sides $[U_{1,2}]$. In practice, notice that we use the following form of the functions $u_{1,2}$, to compute (5.40)

$$
\begin{cases}
u_1(\xi) = S_1(\xi) - \left[\sum_{z \in Z_1} \int_{\Gamma_0} DM(\xi,\zeta) \cdot \frac{w(z)}{z-\zeta} d\zeta \right. \\
\qquad\qquad\qquad \left. + \sum_{z \in Z_2} \int_{\Gamma_0} TM(\xi,\zeta) \cdot \frac{w(z)}{z-\zeta} d\zeta \right] \\
u_2(\xi) = S_2(\xi) - \left[\sum_{z \in Z_1} \int_{\Gamma_0} TM(\xi,\zeta) \cdot \frac{w(z)}{z-\zeta} d\zeta \right. \\
\qquad\qquad\qquad \left. + \sum_{z \in Z_2} \int_{\Gamma_0} DM(\xi,\zeta) \cdot \frac{w(z)}{z-\zeta} d\zeta \right].
\end{cases}
\tag{5.45}
$$

We have implemented these formulas in two ways.

In the first one, all the integrals occuring in (5.37) have been computed by hand with the help of elementary special functions. The calculations, though very tedious, are tractable when the Galerkin basis is given by (5.30). They are not given here.

In the second one, the discrete system is computed starting from (5.35). Each integral on the path Γ_0 (Fig. 2.5) is computed numerically. However, the numerical experiments have proved that this computation should be extremely accurate. The principle of the deformation of the path Γ_0 onto the imaginary axis $\tilde{\Gamma}_0 = \{iy, y \in \mathbb{R}\}$ is still used, as in (5.36), but at a numerical level, in the following way.

Consider a meromorphic function f in $\mathbb{C}\backslash]-\infty, -1]$ well behaved at infinity. We note $(z_m)_{1 \leq m \leq \text{npol}}$ the poles of the function f, $\text{Res}(f, z_m)$ the residue, $\text{sign}(z_m, \Gamma_0) = \pm 1$ if $\pm \text{Im } z_m > 0$, and D_0 the domain

$$D_0 = \left\{ z \in \mathbb{C} / (\text{Re } z > 0, \text{Im } z > 0) \text{ or } (\text{Re } z < 0, \text{Im } z < 0) \right\} \qquad (5.46)$$

The residue formula yields

$$\int_{\Gamma_0} f(\zeta)d\zeta = i \int_{\mathbb{R}} f(iy)dy + (2i\pi) \sum_{m=1}^{\text{npol}} \text{sign}(z_m) \mathbf{1}(z_m \in D_0) \text{Res}(f, z_m).$$
$$(5.47)$$

where the integral on $\mathbb{R}$ of $f(iy)$ is approximated by a progressive Paterson rule, along the recommandations in [E]. In practice, we use the routine D01AMF of the library NAG, which performs such a procedure with a control on the error. The formula (5.47) is applied to each integral occuring in the matrix

$$[D_{\ell k}]_{ij} = \int_{\Gamma_0} DM_{ij}(b_\ell, \zeta) \cdot \varphi_k(\zeta)d\zeta, \quad [T_{\ell k}]_{ij} = \int_{\Gamma_0} TM_{ij}(b_\ell, \zeta) \cdot \varphi_k(\zeta)d\zeta$$
$$(5.48)$$

and to the integrals of the right-hand side $[U_{\alpha,i}]_\ell = u_{\alpha,i}(b_\ell)$ (see (5.45)). The resolution of the linear systems (5.28) is performed by the standard Gauss method (routine F04ADF of the NAG library). Then, $[X_1]$, $[X_2]$ are deduced.

c) Evaluation of the spectral function As in the computation of the poles, the recursive formula leads to a small number $i_{\max}$ of generations $\mathcal{I}_i$ of iterates when φ is large (say, $\frac{\pi}{2} < \varphi < \pi$). On the contrary, the more φ is small, the more $i_{\max}$ is large. In practice, we take as the domain $\mathcal{D}$ of admissibility of the direct formula either $\mathcal{D} = \Omega_L^-$, or $\mathcal{D} = \{\xi, \text{Im } \xi < 0, \text{Re } \xi > x_D\}$ where x_D is a fixed parameter $0 < x_D < 1$. Thus, we evaluate $\Sigma_{1,2}(\xi)$ by (Fig. 5.4)

$$\begin{cases} \bullet & \xi \in \mathcal{D}, \ \Sigma_{1,2}(\xi) = y_{1,2}(\xi) + X_{1,2}(\xi), \\ \bullet & \xi \notin \mathcal{D}, \ \Sigma_{1,2}(\xi) \text{ is computed by (5.43).} \end{cases} \qquad (5.49)$$

In order to limit the number of computations required by the recursive formula, we use the following remark.

Let θ_{obs} be an angle of observation. We have to compute $\Sigma_{1,3}(\xi_1)$ and $\Sigma_{2,3}(\xi_2)$ where $\xi_1 = \nu_0 \cos\theta_{obs}$, $\xi_2 = \nu_0 \cos(\varphi - \theta_{obs})$. We note $\psi_j \in \mathcal{D}$, such that $\xi_j = \nu_0 \cos\psi_j$. There are three cases.

- **only the face 1 is visible**

This case corresponds to $\theta_{obs} \geq \varphi$. We have $\psi_1 = \theta_{obs}$, $\xi_1 \in \Omega_0^-$ and $\psi_2 = \psi_1 - \varphi$, i.e. $\xi_2 = T_0^{-1}(\xi_1)$. Therefore we evaluate directly the amplitude of the diffracted wave $A(\theta_{obs})$ by

$$\Sigma_{1,3}(\xi_1) + \Sigma_{2,3}(\xi_2) = g_1(\xi_1) + \sum_{*\in\{L,T\}} M_*(\xi_1)\Sigma_{2,3}(T_*^{-1}(\xi_1)). \qquad (5.50)$$

- **only the face 2 is visible**

This case corresponds to $\theta_{obs} \leq 0$ We have $\psi_2 = \varphi - \theta_{obs}$, $\psi_1 = |\theta_{obs}|$, i.e. $\xi_1 = T_0^{-1}(\xi_2)$. Therefore the amplitude of the diffracted wave $A(\theta_{obs})$ is given by

$$\Sigma_{1,3}(\xi_1) + \Sigma_{2,3}(\xi_2) = g_2(\xi_2) + \sum_{*\in\{L,T\}} M_*(\xi_2)\Sigma_{1,3}(T_*^{-1}(\xi_2)). \qquad (5.51)$$

- **the two faces are visible**

We have $0 < \theta_{obs} < \varphi$. We have to compute in this case $\Sigma_1(\xi_1)$ and $\Sigma_2(\xi_2)$. If $\varphi \leq \frac{\pi}{2}$, the two points ξ_1, ξ_2 are in the right lower quadrant $\operatorname{Im}\xi < 0$, $\operatorname{Re}\xi > 0$. Therefore they are very close to the domain $\mathcal{D}_0$. If $\frac{\pi}{2} < \varphi < \pi$, the points $T_*^{-1}(\xi)$ enter very quickly in $\mathcal{D}_0$. In all cases, there are very few iterates.

Finally, the recurrence relation (5.44) has been simulated in two different ways. The first is the classical one. We pile up the successive iterates of a point ξ in a stack. A point in the stack is called a *terminal point* if it is located in $\mathcal{D}_0$. It generates no other points. When all the points of the tree have been computed, the contribution of each of them to $\Sigma_{1,2}(\xi)$ is cumulated backward. This algorithm performs correctly up to angle $\varphi \sim 30^\circ$.

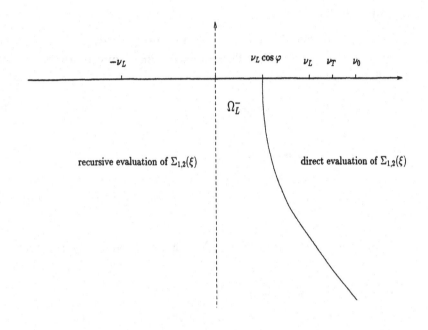

Fig. 5.4. The domain $\mathcal{D}$ for the direct evaluation of the spectral function

For smaller angles ($\varphi \sim 30^\circ$), instabilities in the computation appear for observation angles θ_{obs} close to grazing angles, ($\theta_{obs} \sim -(\pi - \varphi)$ or $\theta_{obs} \sim \pi$. The numerical experiences have proved that in these cases (say $\xi = \nu_0 \cos\theta_{obs} < -\nu_T$), the points of each generation $\mathcal{Z}_i$ of the iterates form clusters, and that the contribution to the spectral function of these clusters remain bounded. This is not the case when the contribution of each term of $\mathcal{Z}_i$ is cumulated separetely : the product of the transfer matrices give rise to an amplification and the final cancellations are not accurate enough. To take in account this phenomenon, we simulate (5.44) in the following way. A point ξ of the half-plane Im $\xi < 0$ is called a *terminal point* when $\Sigma_{1,2}(\xi)$ should be computed without splitting the contribution of the generation $\mathcal{Z}_i$. Let $i_{\max}$ be the total number of generations, n_i the number of points of $\mathcal{Z}_i$, $\xi_{ij} = T_*^{-1}(\tilde{\xi})$ the point j of the generation $\mathcal{Z}_i$, $M_{ij} = M_*(\tilde{\xi})$ the corresponding transfer matrix. We rewrite (5.44) as

$$\Sigma_\alpha(\xi) = g_\alpha(\xi) + \sum_{i=1}^{i_{max}-1} \sum_{j=1}^{n_i} \overline{M}_{ij} g_{\alpha(i)}(\xi_{ij}) + \sum_{j=1}^{n_{i_{max}}} \overline{M}_{i\,max,j}\; \Sigma_{\alpha(i_{max})}(\xi_{i_{max},j})$$

$$(5.52)$$

where $\alpha = \alpha(0) = 1$ or 2, $\alpha(i+1) = 1$ (resp. 2) if $\alpha(i) = 2$ (resp. 1), $\overline{M}_{ij} = M_{1,j'(i,j,1)} M_{2,j'(i,j,2)} \ldots M_{ij}$, and j' (i,j,i') is the index in the generation i' of the ancestror of ξ_{ij}. The contribution of each generation is evaluated separately before being cumulated in $\Sigma_\alpha(\xi)$. Practically, we take as the domain $\mathcal{D}_t$ of the terminal points the points (x,y) ootside of the ellipse

$$\frac{x^2}{1+\varepsilon^2} + \frac{y^2}{\varepsilon^2} > \nu_t^2, \qquad y < 0 \qquad (5.53)$$

where $\varepsilon > 0$ is a fixed parameter. This algorithm allows to reach angles of wedge reaching $\varphi \sim 20°$.

d) Implementation

This algorithm has been implemented in FORTRAN77. The time computation for a complete diagram (see Sect.6) on a SPARC ULTRA 1 varies from a few seconds ($\varphi > 60°$) to a few minutes ($\varphi < 30°$) in the version using the explicit implementation of the integrals, (see Sect.5.6.b)). The version where the integrals are computed numerically uses substantially more CPU, ranging from a few minutes ($\varphi > 60°$) to a few hours ($\varphi < 30°$).

6. Numerical Results

6.1 Introduction

In this section, we present the diagrams of diffraction obtained in the particular case of a wedge in dural immersed in water. The two cases considered are firstly the one of an incident plane wave in the fluid, (Fig 2.1), secondly the one of an incident Scholte-Stoneley incident wave along the face 1 of the wedge, (Fig 5.1). We compute the logarithm $D(\theta_{obs})$, $\theta_{obs} \in] - (\pi - \varphi), \pi[$ of the amplitude $A(\theta_{obs})$, (Sect.2.7), given by

$$D(\theta_{obs}) = \log_{10} |\gamma_1(\nu_0 \cos(\theta_{obs}) + i0^-) + \gamma_2(\nu_0 \cos(\varphi - \theta_{obs}) + i0^-)| \quad (6.1)$$

This function is proportional to the amplitude of the diffracted wave measured in debye. The ratio of the densities is

$$\rho = \rho_f / \rho_s = 2.8 \quad (6.2)$$

The numerical values of the wave velocities c_L, c_T, c_0 are

$$c_L = 6370 ms^{-1}, \quad c_T = 3130 ms^{-1}, \quad c_0 = 1470 ms^{-1}, \quad (6.3)$$

Therefore, the dimensionless coefficients $\nu_* = c_L/c_*$, $* \in L, T, 0$ are

$$\nu_L = 1, \quad \nu_T = 2.035144, \quad \nu_0 = 4.333333 \quad (6.4)$$

The unique root of the equation $R(\xi) = 0$, (Sect.3.2), in the interval $[\nu_T, +\infty[$, is the coefficient of the Rayleigh wave. Its numerical value is

$$\nu_R = 2.1797 \quad (6.5)$$

Similarly, (cf Lemma 3.3), the coefficient ν_S of the Scholte-Stoneley wave is given by $\nu_S/\nu_0 = 1.00195148$, i.e.

$$\nu_S = 4.341790 \quad (6.6)$$

In addition, the vector $w_S \in \mathbb{C}^3$ occuring in the data of the incident Scholte-Stoneley wave is the one given by (5.13).

The coupling of the elasticity equations and of the wave equation along a plane interface, (Sect.3.2), generates the so-called *critical angles*. These angles correspond to the branch points $-\nu_L$, $-\nu_T$ and to the proximity to the real axis of the Rayleigh root $-\nu_R(\rho)$, (3.75). The critical angles θ_*, $* \in \{L, T, R\}$ are computed by

$$\nu_0 \cos \theta_* = \nu_* \tag{6.7}$$

The observation angles corresponding to the critical angles are $\theta_{obs} = \pi - \theta_*$. In the present case (dural/water), we have

$$\theta_L = 76.66°, \quad \theta_T = 61.98°, \quad \theta_R = 59.79° \tag{6.8}$$

The standard Galerkin basis is the one given by (5.30)

$$\varphi_k(\zeta) = d_k/(\zeta + e_k), \quad 1 \le k \le N \tag{6.9}$$

where e_k follows an exponential relation $e_k = p + \varepsilon(10^{\frac{k-1}{h}} - 1)$. Typical values of the parameters p, ε, h, N are

$$p = 1, \quad \varepsilon = 0.05, \quad h = 4., \quad N = 20 \tag{6.10}$$

The collocation points b_l are chosen as $b_l = e_l - 0.1i$

A lot of numerical experiments have been done with slight modifications of these choices.

Though the basis (5.32) seems interesting, it has not yet given better results than the standard one. In particular, the bad behaviour of the direct evaluation (5.42) at the points of the semi-axis $]-\infty, -1] + i0^-$ when the standard basis is used, is not improved by the use of (5.32). Thus, the solution to this problem remains, until now, the use of the recursive formula (5.43) with the tricks of Sect.5.6 . Another problem is that the number of generations of poles occuring in $y_{1,2}(\xi)$ grows exponentially when the angle of the wedge gets smaller, (Sect.5.2-5.3). This gives rise to a prohibitive time computation and to inperfect numerical cancellations in the computation of $y_{1,2}(\xi)$ or of the right hand-side $u_{1,2}(\xi)$ in (5.21).

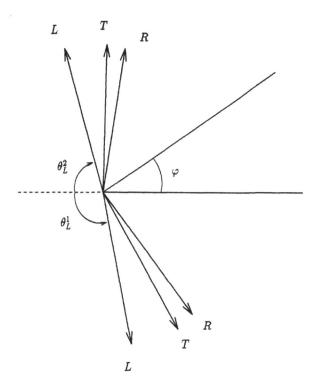

Fig. 6.1. The critical angles for the two faces of the wedge

We have thus used a slight modification of the previously described algorithm. We limit a priori the number of poles to, say, four generations. Consequently we should take in account the forgotten poles in the X-part of the spectral function. Of course, this X-part is no longer holomorphic in $\mathbb{C}-]-\infty,-1]$. We decompose now $X_{1,2}$ on a basis of type (5.30), but with the points e_k distributed on the two semi-axis displayed on Fig 6.2. Such a basis will be called a *wedge-basis* in the sequel. This kind of basis has given very good results, especially for the small angles.

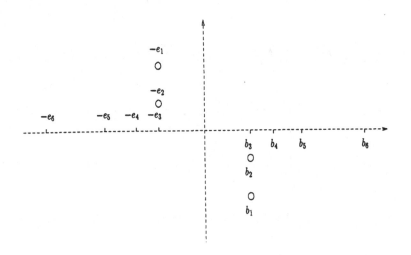

Fig. 6.2. The Galerkin points $-e_k$ and the collocation points b_k distributed on two semi-axis

6.2 Presentation of the Results

We display here the diagamms of diffraction for wedges with an angle $\varphi = 150°, 120°, 90°, 60°, 45°, 25°$. For each of these angles, we consider two cases:

- **Incident Scholte-Stoneley wave**

We display both the diagram of diffraction, and the poles generated by ν_S.

- **Incident wave in the fluid**

We display successively the diagram of diffraction for three angles of incidence. Recall that the incidence angles are measured with respect to the first face of the wedge, (Fig 2.1). The first one is an incidence angle of $\theta_{in} = 180° - 0.5\varphi$, corresponding to an incident wave symmetric with respect to the wedge. The diagram is therefore symmetric. The second incident angle is choosed as $\theta_{in} = 70°$. It lies between the two critical angles θ_L, θ_T

of the plane interface. Therefore, the poles of type T generated by the incident wave, are located on the real axis, giving rise to numerous peaks in the diagram. Conversely, the poles of type L turn around the origin in the upper complex plane. Finally, for $\theta_{in} = 45°$, we have a total reflection on the face 1, because, θ_{in} is smaller than the first critical angle θ_T .

For all wedges, the observation angles corresponding to the critical angles θ_L, θ_T, θ_R are, for the face 1,

$$\theta_L^1 = 103.35°, \quad \theta_T^1 = 118.01°, \quad \theta_R^1 = 120.20° \qquad (6.11)$$

For the face 2, the critical angles are $\theta_*^2 = \varphi - \theta_*^1$. On all the diagrams, we see clearly the bifurcation in the behaviour of the diffracted wave at these angles. In most of the cases, the effect of the Rayleigh angle masks the one of the transversal angle. Moreover, an important loss of power in the diffracted wave occurs frequently in the immediate neighborhood of the Rayleigh angle. The more the angle is small, the more they are poles. For $\varphi = 45°$ and $\varphi = 25°$, the large number of peaks in the diagram is due to the poles of type 0 on the real axis. In the Stoneley case, the maximum of the signal at $\theta_{obs} = 0$ corresponds simply to the incident Scholte-Stoneley wave, whereas in the case of an incident wave in the fluid, we observe two peaks due to the direct and reflected wave.

In all computations, a large number of tests have been performed, in order to check the invariance of the diagram when the Galerkin basis or the location of the collocation points change. All the diagrams have been computed with an increment of $1°$ in the range $\theta_{obs} \in] - (\pi - \varphi), \pi[$.

6.2.1 Wedge of Angle $\varphi = 150°$

The results are displayed on Fig. 6.3 to Fig. 6.8. The critical angles for the second face are

$$\theta_L^2 = 46.65°, \quad \theta_T^2 = 31.99°, \quad \theta_R^2 = 29.80° \qquad (6.12)$$

In the Stoneley case, there are only two generations of poles. The first one is $\mathcal{Z}_0 = \{\nu_S\}$ and the second one is composed by the 3 points on the left of the Fig 6.4 . In the incident volume wave, observe the analogy between the diagrams for a Scholte-Stoneley incidence wave and for the incident wave with $\theta_{in} = 45°$ in the range $\theta_{obs} \in [100°, 180°[$.

6.2.2 Wedge of Angle $\varphi = 120°$

The results are displayed on Fig. 6.9 to Fig. 6.14. The critical angles for the face 2 are

$$\theta_L^2 = 16.65°, \quad \theta_T^2 = 1.99°, \quad \theta_R^2 = -0.2° \tag{6.13}$$

For the Stoneley incidence, the effect of the angles θ_T^2 and θ_R^2 are masked by the incident wave. This is no longer the case at $\theta_{in} = 70°$ where the effect of these two angles are apparent. Note also the loss of power in the signal after $\theta_{obs} = 20°$, which does not correspond to any critical angle.

6.2.3 Wedge of Angle $\varphi = 90°$

The results are displayed on Fig. 6.15 to Fig. 6.20. The critical angles for the face 2 are

$$\theta_L^2 = -13.35°, \quad \theta_T^2 = -28.01°, \quad \theta_R^2 = -30.2° \tag{6.14}$$

For an incident volume wave at $\theta_{in} = 70°$, we observe a peak between the direct and reflected wave, due to the reemission in the fluid of a transverse wave. This wave is due to the pole 'T' on the real axis, which is the iterate of $\nu_0 \cos\theta_{in}$.

6.2.4 Wedge of Angle $\varphi = 60°$

The results are displayed on Fig. 6.21 to Fig. 6.26. The critical angles for the face 2 are

$$\theta_L^2 = -43.35°, \quad \theta_T^2 = -58.01°, \quad \theta_R^2 = -60.2° \tag{6.15}$$

In the Stoneley case, we see that $\mathcal{Z}_2$ is the last complete generation of poles.

6.2.5 Wedge of Angle $\varphi = 45°$

The results are displayed on Fig. 6.27 to Fig. 6.32. The critical angles for the face 2 are

$$\theta_L^2 = -58.35°, \quad \theta_T^2 = -73.01°, \quad \theta_R^2 = -75.20° \tag{6.16}$$

Observe the large loss of power at the angle θ_L^1. Moreover, the effect of the angles θ_T^2 and θ_R^2 are distinguishable on the left of the diagram.

6.2.6 Wedge of Angle $\varphi = 25°$

The results are displayed on Fig. 6.33 to Fig. 6.38. The critical angles for the face 2 are

$$\theta_L^2 = -78.35°, \quad \theta_T^2 = -93.01°, \quad \theta_R^2 = -95.20° \qquad (6.17)$$

They are 8 generations of poles in the case of a Stoneley incident wave. The number of poles in y_1 is 222 and 469 in y_2. The number of generation of poles taken in account in the y-part should be limited to 2 or 3. The computations for this angle are the most difficult of the series. We need to use the second interpretation (5.52) of the recursive formula (3.135). Otherwise, numerical instabilities do appear at the extremities of the diagram.

6.3 Iterates Generated by the Recursive Formula

We display three examples of the iterates generated by the use of the recursive formula (3.135) in the case of a wedge of angle $\varphi = 25°$ with incident volume wave at $\theta_{in} = 70°$. Let $\xi = \nu_0 \cos(\theta_{obs})$ be an observation point corresponding to the face 1 of the wedge, we plot the points in the half-lower plane generated by ξ by the recurrence formula (5.44). In addition are plotted, the ellipse (5.53) with parameter $\varepsilon = 0.5$ and the hyperbola $\partial\mathcal{D} = \partial\Omega_L^-$, (cf 5.49), which determinates the switch between the recursive and the direct evaluation.

6.3.1 $\theta_{obs} = 140°$

We have $\nu_0 \cos\theta_{obs} < -\nu_T 1 + (\varepsilon^2)^{1/2}$. Therefore, the L and T -iterates are in the domain Im $\xi < 0$. In this case, these iterates form clustered generations (cf Sect.5.6) that are displayed on Fig. 6.39.

6.3.2 $\theta_{obs} = 110°$

We have $-\nu_T(1 + \varepsilon^2)^{1/2} < \nu_0 \cos\theta_{obs} < \nu_L$. In this case, the T-iterate of $\nu_0 \cos\theta_{obs}$ remains on the real axis, whereas its L-iterate is in Im $\xi < 0$. The subsequent generations of iterates are not clustered, (Fig. 6.40).

6.3.3 $\theta_{obs} = 95°$

We have $-\nu_L < \nu_0 \cos\theta_{obs} < 0..$ All the iterates of $\nu_0 \cos\theta_{obs}$ remain on the real axis, (Fig. 6.41).

6.4 Numerical Accuracy of the Direct Evaluation

We test here the accuracy of the numerical computation of the X-part of the spectral function in the Stoneley case, for a wedge of $\varphi = 60°$, $\varphi = 45°$, $\varphi = 25°$. We plot along the semi-axis $\xi = \rho e^{-i\theta}$ the function (Fig. 6.42, 6.43, 6.44)

$$\varrho \mapsto \log|E_1 + E_2| \tag{6.18}$$

where E_1 and E_2 are the relative errors

$$E_1 = |DM.\Sigma_1 + TM.\Sigma_2|/|DM.y_1 + TM.y_2| \tag{6.19}$$

$$E_2 = |TM.\Sigma_1 + DM.\Sigma_2|/|DM.y_2 + TM.y_1| \tag{6.20}$$

We see the location of the collocation points on the semi-axis $\theta = 1°$. The accuracy is acceptable on the semi-axis $\theta = 45°$, $\theta = 90°$. It becomes completely lost at the angles $\theta = 135°, \theta = 179°$. Therefore, the recursive computation of the spectral function is effectively needed in these last cases.

6.5 Numerical Diagrams of Diffraction

We display on the following pages the numerical diagrams of diffraction of a dural wedge immersed in water. The angle of the wedge is $\varphi = 150°, 120°, 90°, 60°, 45°, 25°$. We refer to Sect.6.2 for detailed comments.

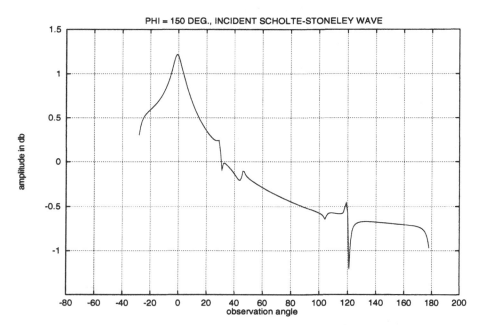

Fig. 6.3. Diagram of diffraction of a dural wedge of $\varphi = 150°$ immersed in water - Incident Scholte-Stoneley wave

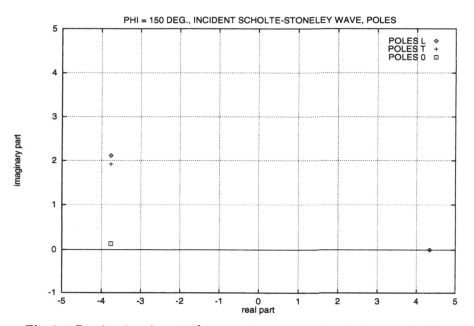

Fig. 6.4. Dural wedge of $\varphi = 150°$ immersed in water : poles of reflection - Incident Scholte-Stoneley wave

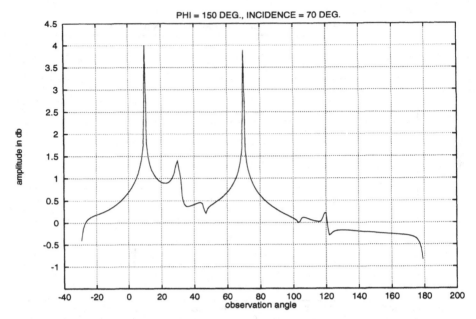

Fig. 6.5. Diagram of diffraction of a dural wedge of $\varphi = 150°$ immersed in water - Incident volume wave, $\theta_{in} = 70°$

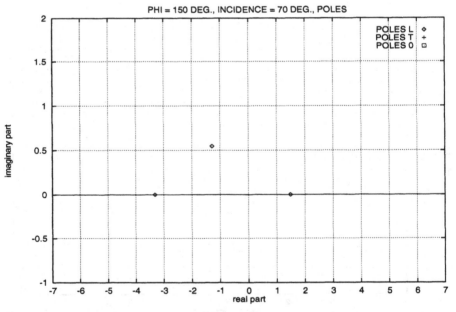

Fig. 6.6. Diagram of diffraction of a dural wedge of $\varphi = 150°$ immersed in water : poles of reflection - Incident volume wave, $\theta_{in} = 70°$

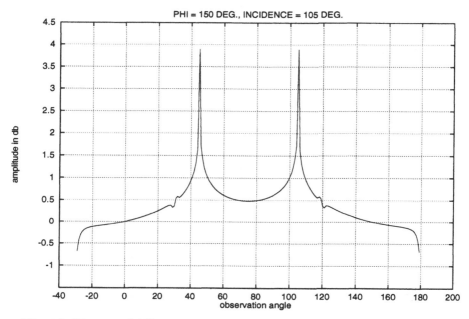

Fig. 6.7. Diagram of diffraction of a dural wedge of $\varphi = 150°$ immersed in water - Incident volume wave, $\theta_{in} = 105°$

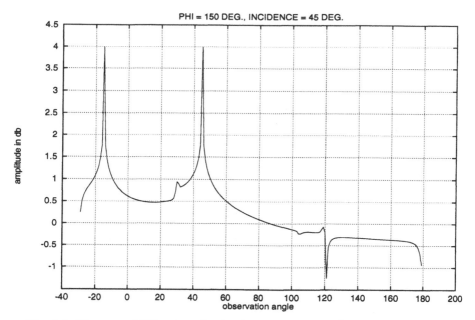

Fig. 6.8. Diagram of diffraction of a dural wedge of $\varphi = 150°$ immersed in water - Incident volume wave, $\theta_{in} = 45°$

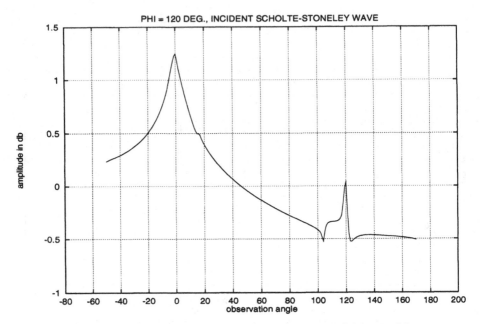

Fig. 6.9. Diagram of diffraction of a dural wedge of $\varphi = 120°$ immersed in water - Incident Scholte-Stoneley wave

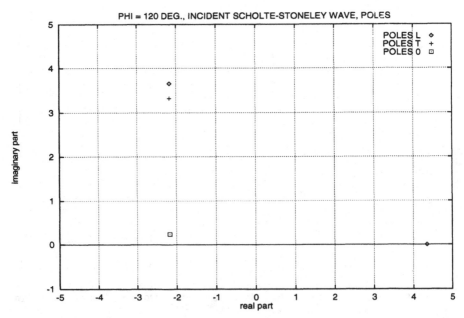

Fig. 6.10. Dural wedge of $\varphi = 120°$ immersed in water : poles of reflection - Incident Scholte-Stoneley wave

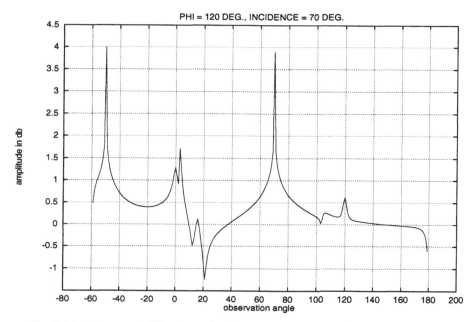

Fig. 6.11. Diagram of diffraction of a dural wedge of $\varphi = 120°$ immersed in water - Incident volume wave, $\theta_{in} = 70°$

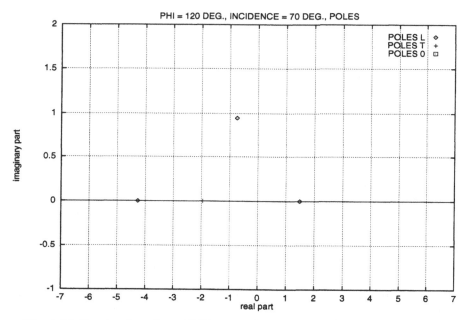

Fig. 6.12. Dural wedge of $\varphi = 120°$ immersed in water : poles of reflection - Incident volume wave, $\theta_{in} = 70°$

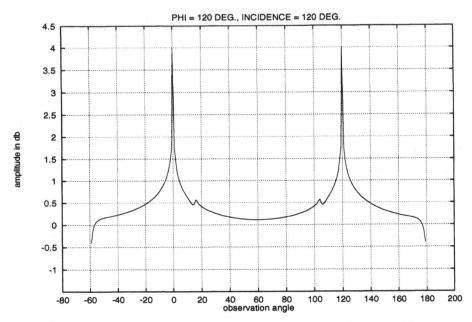

Fig. 6.13. Diagram of diffraction of a dural wedge of $\varphi = 120°$ immersed in water - Incident volume wave, $\theta_{in} = 120°$

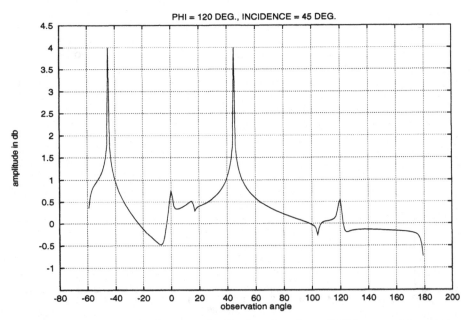

Fig. 6.14. Diagram of diffraction of a dural wedge of $\varphi = 120°$ immersed in water - Incident volume wave, $\theta_{in} = 45°$

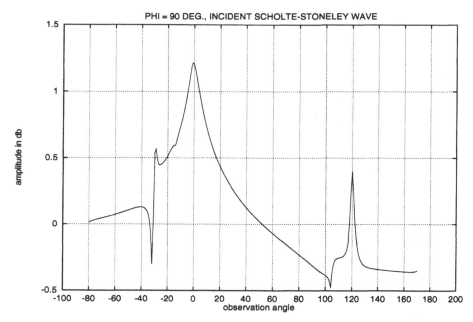

Fig. 6.15. Diagram of diffraction of a dural wedge of $\varphi = 90°$ immersed in water - Incident Scholte-Stoneley wave

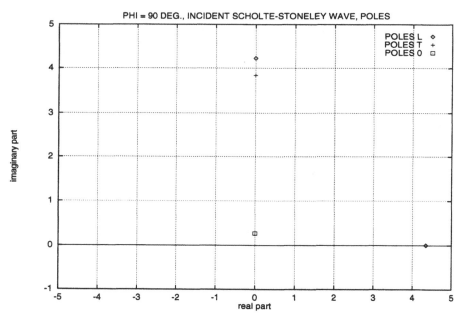

Fig. 6.16. Dural wedge of $\varphi = 90°$ immersed in water : poles of reflection - Incident Scholte-Stoneley wave

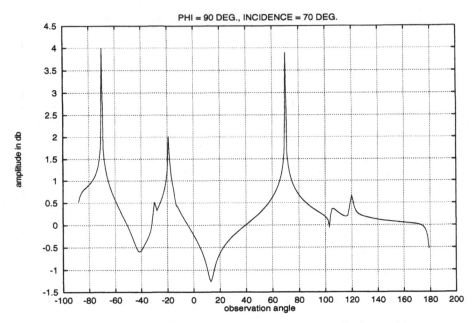

Fig. 6.17. Diagram of diffraction of a dural wedge of $\varphi = 90°$ immersed in water - Incident volume wave, $\theta_{in} = 70°$

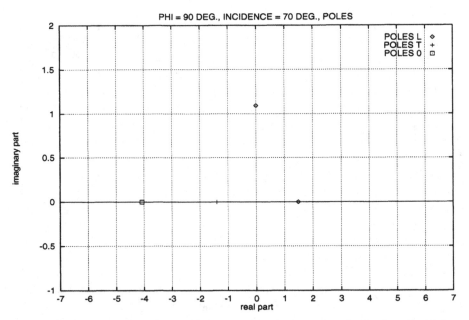

Fig. 6.18. Dural wedge of $\varphi = 90°$ immersed in water : poles of reflection - Incident volume wave, $\theta_{in} = 70°$

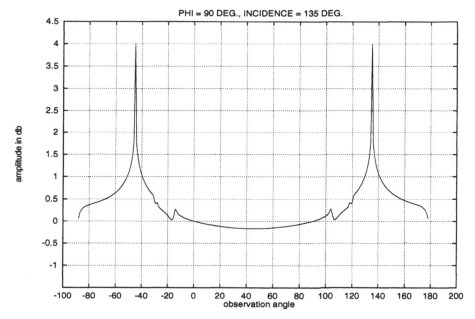

Fig. 6.19. Diagram of diffraction of a dural wedge of $\varphi = 90°$ immersed in water - Incident volume wave, $\theta_{in} = 135°$

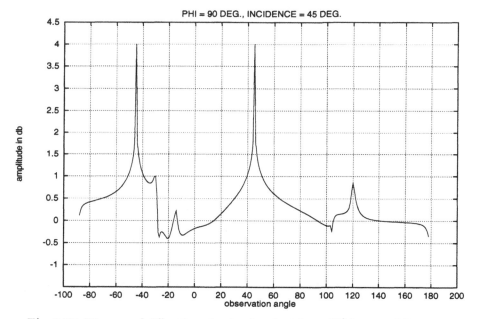

Fig. 6.20. Diagram of diffraction of a dural wedge of $\varphi = 90°$ immersed in water - Incident volume wave, $\theta_{in} = 45°$

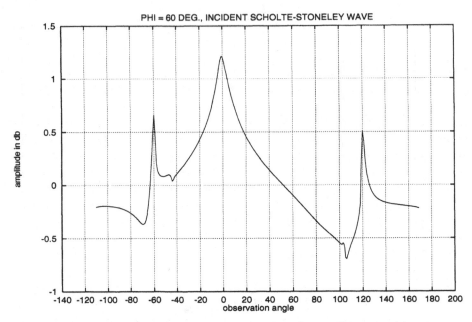

Fig. 6.21. Diagram of diffraction of a dural wedge of $\varphi = 60°$ immersed in water - Incident Scholte-Stoneley wave

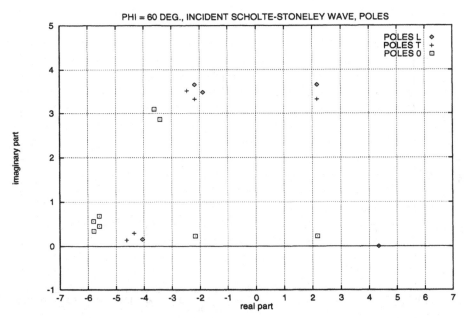

Fig. 6.22. Dural wedge of $\varphi = 60°$ immersed in water : poles of reflection - Incident Scholte-Stoneley wave

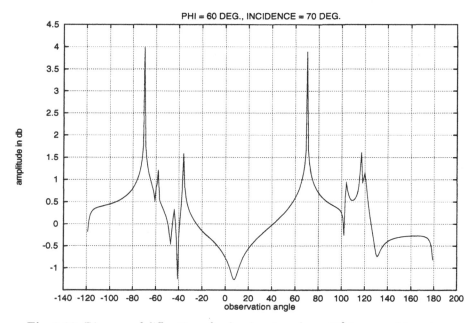

Fig. 6.23. Diagram of diffraction of a dural wedge of $\varphi = 60°$ immersed in water - Incident volume wave, $\theta_{in} = 70°$

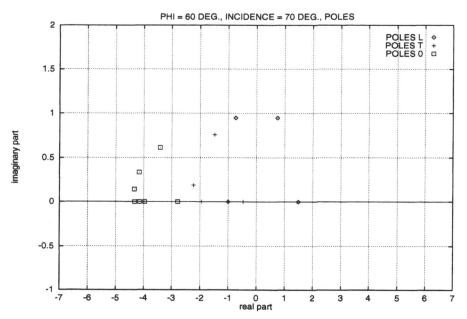

Fig. 6.24. Dural wedge of $\varphi = 60°$ immersed in water : poles of reflection - Incident volume wave, $\theta_{in} = 70°$

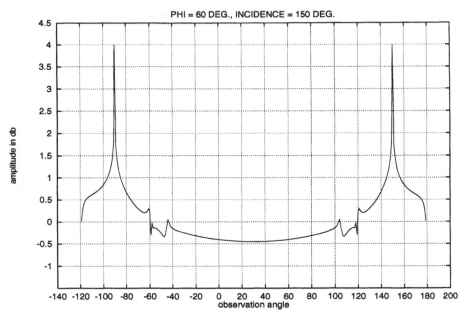

Fig. 6.25. Diagram of diffraction of a dural wedge of $\varphi = 60°$ immersed in water -
Incident volume wave, $\theta_{in} = 150°$

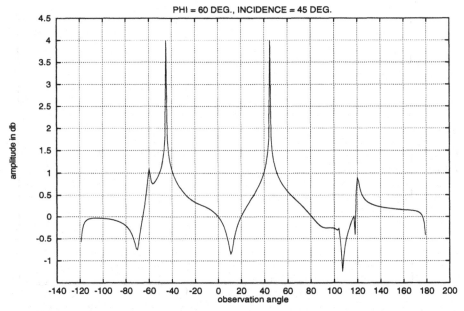

Fig. 6.26. Diagram of diffraction of a dural wedge of $\varphi = 60°$ immersed in water -
Incident volume wave, $\theta_{in} = 45°$

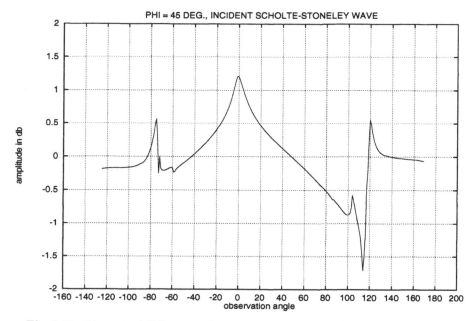

Fig. 6.27. Diagram of diffraction of a dural wedge of $\varphi = 45°$ immersed in water -
Incident Scholte-Stoneley wave

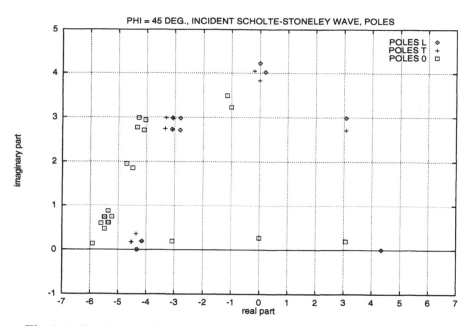

Fig. 6.28. Dural wedge of $\varphi = 45°$ immersed in water : poles of reflection - Incident
Scholte-Stoneley wave

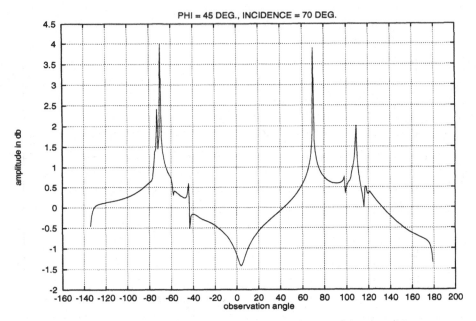

Fig. 6.29. Diagram of diffraction of a dural wedge of $\varphi = 45°$ immersed in water -
Incident volume wave, $\theta_{in} = 70°$

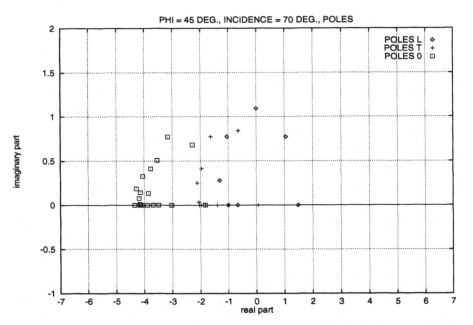

Fig. 6.30. Dural wedge of $\varphi = 45°$ immersed in water : poles of reflection - Incident
volume wave, $\theta_{in} = 70°$

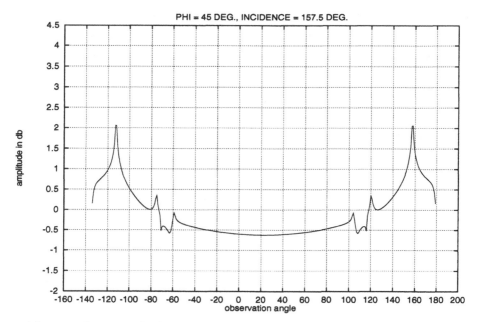

Fig. 6.31. Diagram of diffraction of a dural wedge of $\varphi = 45°$ immersed in water - Incident volume wave, $\theta_{in} = 157.5°$

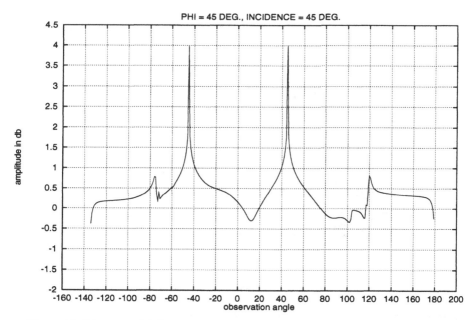

Fig. 6.32. Diagram of diffraction of a dural wedge of $\varphi = 45°$ immersed in water - Incident volume wave, $\theta_{in} = 45°$

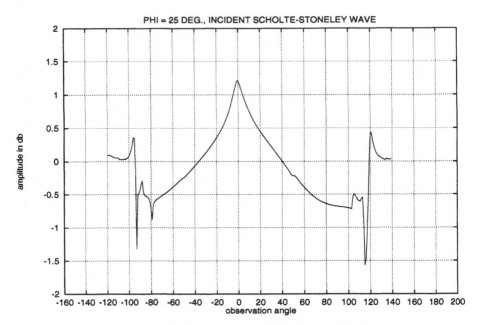

Fig. 6.33. Diagram of diffraction of a dural wedge of $\varphi = 25°$ immersed in water - Incident Scholte-Stoneley wave

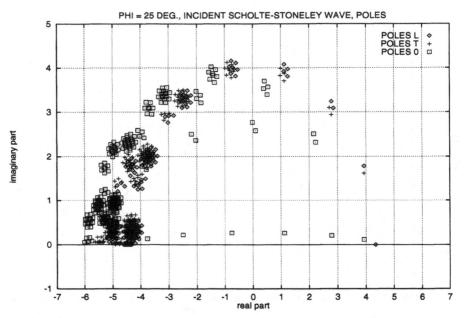

Fig. 6.34. Dural wedge of $\varphi = 25°$ immersed in water : poles of reflection - Incident Scholte-Stoneley wave

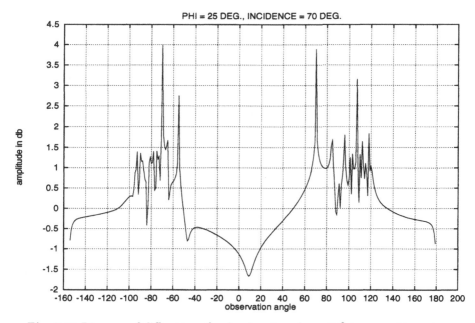

Fig. 6.35 Diagram of diffraction of a dural wedge of $\varphi = 25°$ immersed in water - Incident volume wave, $\theta_{in} = 70°$

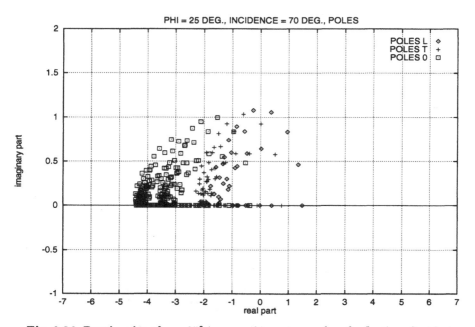

Fig. 6.36. Dural wedge of $\varphi = 25°$ immersed in water : poles of reflection - Incident volume wave, $\theta_{in} = 70°$

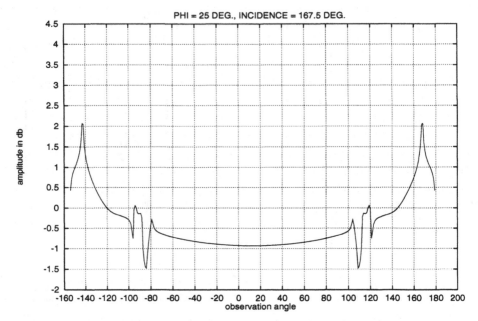

Fig. 6.37. Diagram of diffraction of a dural wedge of $\varphi = 25°$ immersed in water - Incident volume wave, $\theta_{in} = 167.5°$

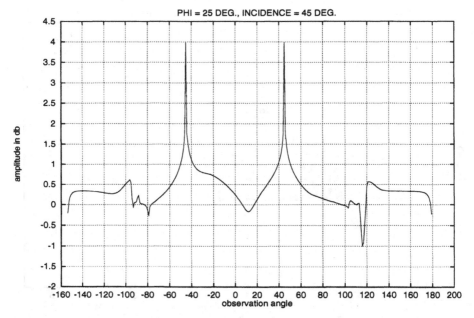

Fig. 6.38. Diagram of diffraction of a dural wedge of $\varphi = 25°$ immersed in water - Incident volume wave, $\theta_{in} = 45°$

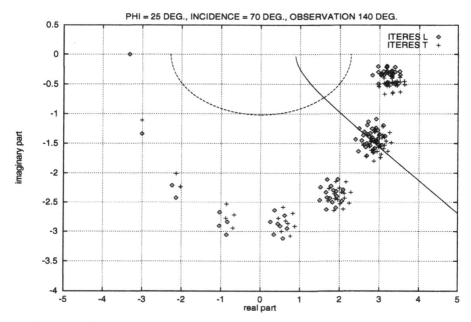

Fig. 6.39. Wedge of angle $\varphi = 25°$, incident wave in the fluid with $\theta_{in} = 70°$ - Iterates of the point $\xi = \nu_0 \cos 140°$.

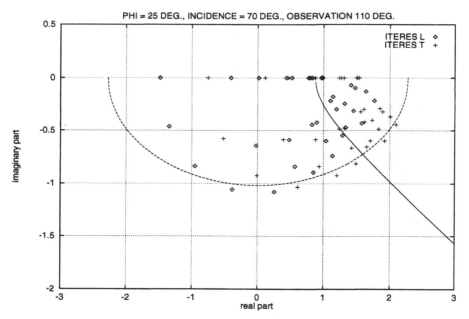

Fig. 6.40. Wedge of angle $\varphi = 25°$, incident wave in the fluid with $\theta_{in} = 70°$ - Iterates of the point $\xi = \nu_0 \cos 110°$.

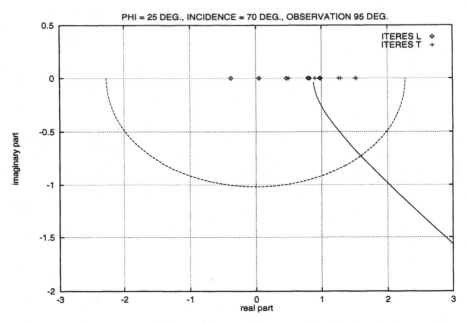

Fig. 6.41. Wedge of angle $\varphi = 25°$, incident wave in the fluid with $\theta_{in} = 70°$ - Iterates of the point $\xi = \nu_0 \cos 95°$.

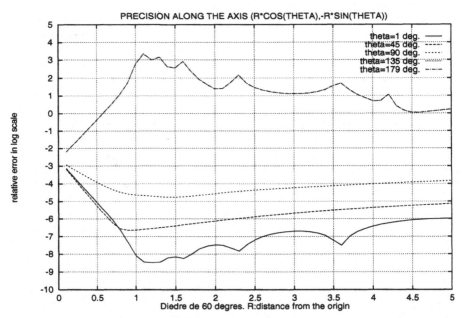

Fig. 6.42. Wedge of angle $\varphi = 60°$, incident Stoneley wave - Relative error along the semi-axis $\xi = \rho e^{-i\theta}$

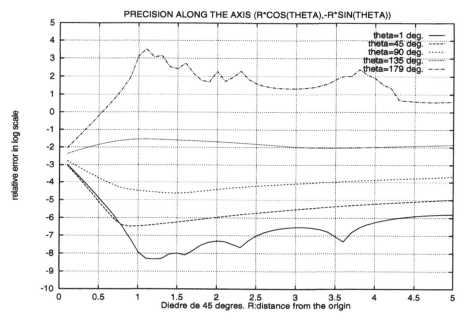

Fig. 6.43. Wedge of angle $\varphi = 45°$, incident Stoneley wave - Relative error along the semi-axis $\xi = \rho e^{-i\theta}$

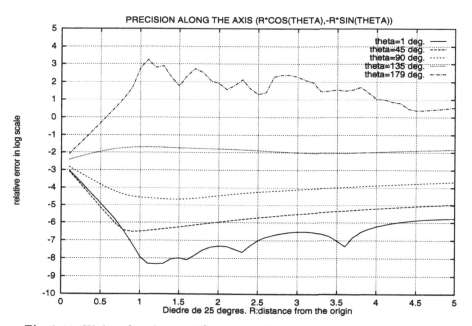

Fig. 6.44. Wedge of angle $\varphi = 25°$, incident Stoneley wave - Relative error along the semi-axis $\xi = \rho e^{-i\theta}$

Appendix

A1. The Hardy Inequality

We use at several places in the text the following result.

Proposition A1. *The kernel operator $1(x \geq 0, y \geq 0)\frac{1}{x+y}$ is bounded on $L^2(\mathbb{R}^+)$.*

Proof. We call A and A_1 the kernel operators defined by

$$A(x,y) = 1(x \geq 0, y \geq 0)\frac{1}{x+y}, \quad A_1(x,y) = 1(x \geq y \geq 0)\frac{1}{x+y}$$

We have $A = A_1 + A_1^*$, and it is sufficient to prove that A_1 is bounded on $L^2(\mathbb{R}^+)$. Due to the estimate

$$|A_1 f(x)| = |\int_0^x \frac{f(y)}{x+y}dy| \leq \frac{1}{x}\int_0^x |f(y)|dy$$

it is sufficient to prove that the operator A_2 defined by

$$A_2 f(x) = \frac{1}{x}\int_0^x f(y)dy = \int_0^1 f(tx)dt$$

is bounded on $L^2(\mathbb{R}^+)$. For any $g \in L^2(\mathbb{R}^+)$, we have

$$(A_2 f, g) = \int_0^1 \left(\int_0^\infty f(tx)\overline{g(x)}dx\right)dt \leq \|g\|M$$

where the constant M is

$$M = \int_0^1 \left(\int_0^\infty |f(tx)|^2 dx\right)^{1/2} dt = \|f\| \int_0^1 \frac{dt}{\sqrt{t}} = 2\|f\|$$

which proves the result. ∎

A2. The Hilbert projector

Recall that we call H^+ the Hilbert space of the functions $f(\xi)$, analytic in the lower half-plane and uniformly L^2 on the horizontal lines, $i.e.$ (cf Sect.3.3)

$$\sup_{c>0} \int_{\mathbf{R}} |f(\xi - ic)|^2 d\xi < +\infty$$

equipped with the norm

$$\|f\| = \left(\int_{\mathbf{R}} |f(\xi)|^2 d\xi \right)^{1/2}$$

Definition A2.1. *The Hilbert projector is the application* $H : L^2(\mathbf{R}) \to H^+$ *defined for* $g \in L^2(\mathbf{R})$ *by*

$$Hg(\xi) = \frac{1}{2\pi} \int_{\mathbf{R}} \frac{g(\xi)}{\xi - \zeta} d\zeta$$

We check easily that if the function $g = \mathcal{F}f$ is the Fourier-Plancherel transform of $f \in L^2(\mathbf{R})$, then Hg is defined by $Hg(\xi) = \mathcal{F}(f(x)\mathbf{1}(x \geq 0))(\xi)$. Therefore $\|Hg\|_{H^+} = \|f\|_{L^2(\mathbf{R}^+)} \leq \|f\|_{L^2(\mathbf{R})} = \|g\|_{L^2(\mathbf{R})}$. This proves the following

Proposition A2.2. *The Hilbert projector is bounded from* $L^2(\mathbf{R})$ *onto* H^+.

Subject Index

References

[Be1]   Bernard, J.M.L.: On the diffraction of an electromagnetic skew incident wave by a non perfectly conducting wedge, *Ann. Telecom*, 45, (1990), 30-39

[Be2]   Bernard, J.M.L.: On the time domain scattering by a classical frequency dependent wedge-shaped region in a longdispersive medium, *Ann. Telecom*, 49, (1994), 673-683

[BM]   Bouche, D., Molinet, F.: Méthodes asymptotiques en électromagnétisme, *Mathématiques et Applications, n° 16*, Springer, 1994

[BSU]   Bowman, J.J., Senior, T.B.A, Uslenghi, P.L.: Acoustic and electromagnetic scattering by simple shapes, Hemisphere, 1987

[CT]   Cheeger, J., Taylor, M.: Diffraction by conical singularities, *Comm. Pure App. Math.*, 35, (1982), 275-487

[CL]   Croisille, J.P., Lebeau, G.: Diffraction de l'onde de Scholte par un dièdre élastique immergé: Approximation numérique, *rapport DRET n° 3/93-2543A, Université Paris-Sud, (1995), unpublished*

[DTDL]   Duflo, H., Tinel, A., Duclos, J., Lebeau, G.: Scholte wave diffraction by a dihedral: Study at oblique incidence *J. Acoust. Soc. Am.* 98, 6 (1995), 3493-3500

[DTD]   Duflo, H., Tinel, A., Duclos, J.: Scholte wave diffraction by a triangular groove, *4th Int. Cong. on Sound and Vibration, St Petersburg, (1996)*

[E]   Evans, G.: *Practical Numerical Integration*, Wiley, 1993

[Ga]   Garnir, H.G.: Fonction de Green pour l'opérateur métaharmonique dans un angle ou un dièdre, *Bull. Soc. Roy. Sci. Liège*, (1952), 119-140, 207-231, 328-344

[GL]   Gérard, P., Lebeau, G.: Diffusion d'une onde par un coin, *J. of the A.M.S.*, 6, (1993), 341-423

[K]   Keller, J.B.: Geometrical theory of diffraction, *J. Opt. Soc. America*, 52, (1962), 116-130

[L1]   Lebeau, G.: Propagation des ondes dans les dièdres, *Mémoire de la Soc. Math. de France*, 60, (1995), Suppl. au Bull.de la SMF, 123, fasc. 1

[L2]   Lebeau, G.: Diffraction de l'onde de Scholte par un dièdre élastique immergé, Rapport DRET n° 1/93-2543A, Université Paris-Sud, (1992), unpublished

[L3]   Lebeau, G.: Propagation des ondes dans les variétés à coins, *Ann. Scient. ENS*, 30 (1997),429-497

[Ma1]   Maliuzhinets, G.D.: Excitation reflection and emission of surface waves from a wedge with given face impedances, *Sov. Phys. Dokl*, 3, (1958), 752-753

[Ma2]   Maliuzhinets, G.D.: Inversion formula for the Sommerfeld integral, *Sov. Phys. Dokl*, 3, (1958), 52-56

[PB1]   Piet, J.F., de Billy, M.: Experimental study of the scattering of a compressional wave by an elastic wedge, *J. Appl. Phys.*, 69, 10, (1991)

[PB2]   Piet, J.F., de Billy, M.: Experimental study of the ultrasonic field scattered from an immersed elastic wedge, *Colloque C1, supplément au Journal de Physique III, 2, (1992)*

[Po1]   Poincaré, H.: Sur la polarisation par diffraction, (Part 1), *Acta Mathematica*, 16, (1892), 297-339

[Po2]   Poincaré, H.: Sur la polarisation par diffraction, (Part 2), *Acta Mathematica*, 20, (1896), 313-356

[Som1]   Sommerfeld, A.: Matematische Theorie der Diffraction, *Math. Annalen*, 47, (1896), 317-374

[Som2]   Sommerfeld, A.: *Zeit. fur Math. Phys.*, 56, (1901)

[TD]   Tinel, A., Duclos, J.: Diffraction and conversion of the Scholte-Stoneley wave at the extremity of a solid, *J. Acoust. Soc. Am. 95* 1 (1994), 13-20

Lecture Notes in Mathematics

For information about Vols. 1–1530
please contact your bookseller or Springer-Verlag

4. Lecture Notes are printed by photo-offset from the master-copy delivered in camera-ready form by the authors. Springer-Verlag provides technical instructions for the preparation of manuscripts. Macro packages in T_EX, L^AT_EX2e, $L^AT_EX2.09$ are available from Springer's web-pages at

http://www.springer.de/math/authors/b-tex.html.

Careful preparation of the manuscripts will help keep production time short and ensure satisfactory appearance of the finished book.

The actual production of a Lecture Notes volume takes approximately 12 weeks.

5. Authors receive a total of 50 free copies of their volume, but no royalties. They are entitled to a discount of 33.3% on the price of Springer books purchase for their personal use, if ordering directly from Springer-Verlag.

Commitment to publish is made by letter of intent rather than by signing a formal contract. Springer-Verlag secures the copyright for each volume. Authors are free to reuse material contained in their LNM volumes in later publications: A brief written (or e-mail) request for formal permission is sufficient.

Addresses:

Professor F. Takens, Mathematisch Instituut,
Rijksuniversiteit Groningen, Postbus 800,
9700 AV Groningen, The Netherlands
E-mail: F.Takens@math.rug.nl

Professor B. Teissier, DMI, École Normale Supérieure
45, rue d'Ulm,
F-7500 Paris, France
E-mail: Teissier@ens.fr

Springer-Verlag, Mathematics Editorial, Tiergartenstr. 17,
D-69121 Heidelberg, Germany,
Tel.: *49 (6221) 487-701
Fax: *49 (6221) 487-355
E-mail: lnm@Springer.de